HIGH NONLINEARITY OPTICAL FIBER TECHNOLOGY

HIGH NONLINEARITY OPTICAL FIBER TECHNOLOGY

JU HAN LEE

Nova Science Publishers, Inc.
New York

For permission to use material from this book please contact us:
Telephone 631-231-7269; Fax 631-231-8175
Web Site: http://www.novapublishers.com

LIBRARY OF CONGRESS CATALOGING-IN-PUBLICATION DATA

ISBN : 978-1-60692-674-1

Available upon request

Published by Nova Science Publishers, Inc. ✢ New York

CONTENTS

PREFACE

All-optical signal processing technology has been of high technical interest in the field of fiber-optic communication systems and networks, since it can provide the most powerful way to overcome the optical-to-electrical domain conversion-induced data traffic bottleneck. A variety of all-optical signal processing devices demonstrated to date: for example, wavelength converters, optical demultiplexers, noise rejection filters, amplifiers, clock recovery subsystems, and data regenerators. Those devices are mostly based on the nonlinear optical properties of optical waveguide media such as optical fibers and semiconductor active devices. The two main forms of nonlinear effect used are Kerr nonlinear effects and inelastic nonlinear scattering effects (i.e. Raman and Brillouin scattering). Kerr nonlinear effects including self-phase modulation, cross-phase modulation, and four-wave mixing are due to the response of the bound electrons in a nonlinear optical medium to an intense optical field, whereas Raman and Brillouin scattering effects are caused by the presence of vibrational states of atoms in the optical medium. In this chapter, the ultimate potential of state-of-the-art highly nonlinear optical fiber technologies are reviewed from a perspective of practical implementation of all-optical signal processing devices for fiber optic communication systems. Due to the fact that the ultra-high nonlinearity fiber technologies offer significant advantages in terms of reduced length, reduced power requirements for the realization of a variety of all-optical signal processing devices, the compactness and stability issue of fiber-based optical devices relative to semiconductor-based ones can be significantly improved. This review is focused on two state-of-the-art ultra-high nonlinearity optical fibers such as photonic crystal fiber (PCF) and Bismuth oxide-based step index type nonlinear optical fiber, which are considered to be the most promising fibers among various kinds of nonlinear optical fiber. Also presented are basics on optical fiber nonlinearities to help the readers to understand the main contents. In addition, nonlinearity efficiency comparison results for various types of commercially available, state-of-the-art nonlinear fiber are presented in terms of various definitions of Kerr nonlinearity figure-of-merit to provide critical information with regard to optimum optical fiber structure and material for the best Kerr nonlinearity efficiency.

Chapter 1

INTRODUCTION

The huge demands for the expansion of data transmission capacity in fiber-optic telecommunication systems and related optical networks, mean that the development of all-optical signal processing devices and technologies with which to tackle the performance limitations that otherwise result from the unavoidable conversion from the optical to electrical domains has become of great practical interest. A great number of signal processing approaches and devices based on the nonlinear properties of both optical fibers and semiconductor active devices have been studied and successfully applied to a variety of applications in telecommunication systems. Devices demonstrated to date include amongst others, wavelength converters [1, 2, 3], optical demultiplexers [4, 5], noise rejection filters [6, 7], amplifiers [8, 9], clock recovery subsystems [10], and data regenerators [11]. The two main forms of nonlinear effect used are (a) Kerr nonlinear effects and (b) inelastic nonlinear scattering effects (i.e. Raman and Brillouin scattering). Kerr nonlinear effects including self-phase modulation, cross-phase modulation, and four-wave mixing are due to the response of the bound electrons in a nonlinear optical medium to an intense optical field, while Raman and Brillouin scattering effects are caused by the presence of vibrational states of atoms (phonons) in the optical medium.

Until now, semiconductor-based devices such as semiconductor optical amplifiers (SOAs) [12] and electroabsorption modulators (EAMs) [13] are a preferred choice to optical fibers in the implementation of all-optical signal processing devices due to the significant drawback of optical fibers that is the stability and compact issues associated with its long fiber length. For instance, in order to obtain a π phase shift in the case of using a conventional standard single mode fiber (SMF) as a nonlinear medium, into which is a 100 mW laser beam launched, at least 30 km length of the fiber is required. Even if the 30-km length of the fiber can be wound onto a small size spool, it would not be a good choice from a perspective of compact size device integration.

In this chapter, highly nonlinear optical fiber technologies are reviewed from a viewpoint of practical implementation of all-optical signal processing devices. This review is focused on two state-of-the-art ultrahigh nonlinearity optical fibers such as photonic crystal fiber (PCF) [14] and Bismuth oxide based step index type nonlinear optical fiber [15], which are considered to be the most promising fibers among various kinds of nonlinear optical fiber.

Various nonlinear effects in optical fibers are first reviewed in Section 2. This review is carried out from a positive perspective focusing on all-optical nonlinear signal processing device applications. However, it should be appreciated that these effects can be highly undesirable in other instances. For example, four-wave mixing induced crosstalk between WDM channels [16] and Raman effect induced signal power depletion [17] can significantly degrade the performance of DWDM communication systems.

Section 3 presents a study on nonlinearity efficiency comparison for various types of commercially available, state-of-the-art nonlinear fiber in terms of Kerr nonlinearity figure-of-merit (FOM). A novel definition of optical fiber Kerr nonlinearity FOM including the stimulated Brillouin scattering caused pump beam power limit is introduced and compared with the conventional FOMs. This study provides critical information with regard to optimum optical fiber structure and material for the best Kerr nonlinearity efficiency.

In Section 4 and 5, two state-of-the-art ultra-high nonlinearity fiber technologies such as silica-based photonic crystal fiber (PCF) and Bismuth oxide-based nonlinear optical fiber (Bi-NLF) are investigated as a strong candidate for implementation of a range of nonlinear signal processing devices. Due to the fact that those ultra-high nonlinearity fiber technologies offer significant advantages in terms of reduced length, reduced power requirements for a variety of nonlinear optical devices owing to ultra-high nonlinearity of those fibers 10 ~ 1000 times larger than that of conventional silica-based step index fibers, the compactness and stability issue of fiber-based optical devices relative to semiconductor-based ones can be significantly improved. Various nonlinear signal processing devices achievable with short length of the two ultra-high nonlinearity fibers are also reviewed.

Section 6 presents a conclusion regarding the ultra-high nonlinearity fiber technologies mentioned in the previous sections and discusses their further applications.

Chapter 2

NONLINEARITIES IN OPTICAL FIBER

2.1. WAVE EQUATION

Optical fiber shows a nonlinear response to intense optical fields. This nonlinear response originates from anharmonic motion of bound electrons under the influence of an applied field. The electric dipole induced nonlinear polarization vector $\vec{P}$ can thus be described by the following general relation [18]

$$\vec{P} = \varepsilon_0\left(\chi^{(1)} \cdot \vec{E} + \chi^{(2)} : \vec{E}\vec{E} + \chi^{(3)} \vdots \vec{E}\vec{E}\vec{E} + \cdots\right) \tag{1}$$

where ε_0 is the vacuum permittivity and $\chi^{(j)}(j = 1,2,3,\cdots)$ is jth order susceptibility. $\chi^{(j)}$ is a tensor of rank $j+1$ including the light polarization effects, the effects of which are included through the refractive index n and the attenuation coefficient α. $\chi^{(1)}$ is the linear susceptibility. The second-order susceptibility $\chi^{(2)}$ is responsible for nonlinear effects such as second-harmonic generation and sum-frequency generation. $\chi^{(3)}$ is the third-order susceptibility responsible for phenomena such as third-harmonic generation, four-wave mixing, and the intensity dependence of the refractive index (Kerr effect). In particular, the nonlinear processes, which generate new frequency components, i.e. third-harmonic generation or four-wave mixing requires special efforts to achieve phase matching. One noticeable fact is that inversion symmetric molecule such as SiO_2

has a zero value for $\chi^{(2)}$. As a result optical fibers made of SiO_2 do not generally exhibit second-order nonlinear effects but third-order ones.

For a better understanding of the nonlinear phenomena in optical fiber, it is necessary to consider the theory of electromagnetic wave propagation in dispersive nonlinear media. The propagation of optical fields in optical fiber is governed by Maxwell's equations as follows [18].

$$\nabla \times \vec{E} = -\frac{\partial \vec{B}}{\partial t} \tag{2}$$

$$\nabla \times \vec{H} = \vec{J} + \frac{\partial \vec{D}}{\partial t} \tag{3}$$

$$\nabla \cdot \vec{D} = \rho \tag{4}$$

$$\nabla \cdot \vec{B} = 0 \tag{5}$$

where $\vec{E}$ and $\vec{H}$ represent electric and magnetic field vectors, respectively, and $\vec{D}$ and $\vec{B}$ are the corresponding electric and magnetic flux densities. $\vec{J}$ and ρ represent the current density vector and the magnetic flux density, each but they must be zero because of no existent electromagnetic sources within optical fiber. The flux densities $\vec{D}$ and $\vec{B}$ can be expressed through the following constitutive relations since they originate from the electric and magnetic fields ($\vec{E}$ and $\vec{H}$) propagating inside the optical fiber.

$$\vec{D} = \varepsilon_0 \vec{E} + \vec{P} \tag{6}$$

$$\vec{B} = \mu_0 \vec{H} + \vec{M} \tag{7}$$

where $\vec{M}$ is the induced magnetic polarization vector. In the case of nonmagnetic media like optical fibers $\vec{M} = 0$. $\vec{P}$ is the induced electric polarization vector. The above Maxwell's equations can be converted into the wave equation as below that describes light propagation in optical fiber by taking the curl of Eq. (2) and using Eqs. (3), (6), and (7).

$$\nabla \times \nabla \times \vec{E} = -\frac{1}{c^2}\frac{\partial^2 \vec{E}}{\partial t^2} - \mu_0 \frac{\partial^2 \vec{P}}{\partial t^2} \tag{8}$$

where c is the velocity of light in vacuum and the relation $\mu_0 \varepsilon_0 = 1/c^2$ was used. . The wave equation of Eq. (8) can be further simplified by using the relation $\nabla \times \nabla \times \vec{E} = \nabla(\nabla \cdot \vec{E}) - \nabla^2 \vec{E} = -\nabla^2 \vec{E}$ where $\nabla \cdot \vec{D} = \varepsilon \nabla \cdot \vec{E} = 0$ and ε represents the medium dielectric constant. The equation thus can be written in the following form.

$$\nabla^2 \vec{E} - \frac{1}{c^2}\frac{\partial^2 \vec{E}}{\partial t^2} = \mu_0 \frac{\partial^2 \vec{P}}{\partial t^2} \tag{9}$$

The polarization vector $\vec{P}$ can be expanded into a power series of $\vec{E}$ as follows when $\vec{E}$ is sufficiently weak.

$$\begin{aligned}
\vec{P}(\vec{r},t) &= \vec{P}^{(1)}(\vec{r},t) + \vec{P}^{(2)}(\vec{r},t) + \vec{P}^{(3)}(\vec{r},t) + \cdots \\
&= \varepsilon_0 \int_{-\infty}^{\infty} \chi^{(1)}(\vec{r}-\vec{r},t-t') \cdot \vec{E}(\vec{r},t')d\vec{r}dt' \\
&+ \varepsilon_0 \int_{-\infty}^{\infty} \chi^{(2)}(\vec{r}-\vec{r}_1,t-t_1;\vec{r}-\vec{r}_2,t-t_2) : \vec{E}(\vec{r}_1,t_1)\vec{E}(\vec{r}_2,t_2)d\vec{r}_1dt_1d\vec{r}_2dt_2 \\
&+ \varepsilon_0 \int_{-\infty}^{\infty} \chi^{(3)}(\vec{r}-\vec{r}_1,t-t_1;\vec{r}-\vec{r}_2,t-t_2;\vec{r}-\vec{r}_3,t-t_3) \vdots \vec{E}(\vec{r}_1,t_1)\vec{E}(\vec{r}_2,t_2)\vec{E}(\vec{r}_3,t_3)d\vec{r}_1dt_1d\vec{r}_2dt_2d\vec{r}_3dt_3 \\
&+ \cdots
\end{aligned} \tag{10}$$

where $\chi^{(1)}$ is the linear susceptibility and $\chi^{(n)}$ ($n > 1$) is the nth-order nonlinear susceptibility. If the following two assumptions are introduced: the nonlinear response is instantaneous, in other words, the contribution of molecular vibration to $\chi^{(n)}$ (Raman effects) is neglected, and the medium response is local, Eqs, (10) reduces to Eq. (1) [18].

2.2. Self-Phase Modulation

The most common effect of fiber nonlinearity is known as self-phase modulation (SPM). The nonlinear effect results from the intensity dependence of the refractive index. The nonlinear refractive index in optical fiber can be expressed as [19]

$$n = n_0 + n_2 I \tag{11}$$

where n_0 is the linear refractive index and n_2 is the nonlinear refractive index. I represents the optical intensity of the signal ($I = P_{optical}/A_{eff}$, $P_{optical}$: optical power, A_{eff}: effective mode area). Both n_0 and n_2 are functions of material composition and waveguide structure of the optical fiber. As the signal peak power increases the refractive index of the fiber core also increases.

The main features of SPM can be theoretically understood during the course of solving the wave equation of Eq. (9). The process of solving Eq. (9) requires several assumptions [20]. First, $\vec{P}^{(n)}(\vec{r},t)$ (n > 1) is regarded as a small perturbation to $\vec{P}^{(1)}(\vec{r},t)$. Second, the polarization status of the optical field is maintained along the fiber length. Third, the optical field has a spectral width $\Delta\omega$ much smaller than the vacuum light frequency ω_0 ($\Delta\omega << \omega_0$). Using the slowly varying envelope approximation, the electric field and polarization components can be written in the following forms.

$$\vec{E}(\vec{r},t) = \mathrm{Re}[\hat{E}(\vec{r},t)e^{-i\omega t}] = \mathrm{Re}[\hat{\sigma}E(\vec{r},t)e^{-i\omega t}] \tag{12}$$

$$\vec{P}^{(1)}(\vec{r},t) = \mathrm{Re}[\hat{P}^{(1)}(\vec{r},t)e^{-i\omega t}] \tag{13}$$

$$\vec{P}^{(3)}(\vec{r},t) = \mathrm{Re}[\hat{P}^{(3)}(\vec{r},t)e^{-i\omega t}] \tag{14}$$

where $\hat{\sigma}$ is the polarization unit vector of the light. Note that $P^{(2)}(\vec{r},t)$ term is neglected since optical fibers made of Silica do not generally exhibit second-order nonlinear effects. $E(\vec{r},t)$, $P^{(1)}(\vec{r},t)$, and $P^{(3)}(\vec{r},t)$ are slowly varying functions of time. From Eq. (1)

$$\vec{P}^{(1)}(\vec{r},t) = \varepsilon_0 \chi^{(1)} \vec{E}(\vec{r},t) \tag{15}$$

$$\vec{P}^{(3)}(\vec{r},t) = \varepsilon_0 \chi^{(3)} [\vec{E}(\vec{r},t) \cdot \vec{E}(\vec{r},t)] \vec{E}(\vec{r},t) \tag{16}$$

After $E(\vec{r},t)$ in Eq. (16) is replaced with Eq. (12) $\hat{P}^{(3)}(\vec{r},t)$ is found to have terms at frequencies of ω_0 and $3\omega_0$. Since the third harmonic term can generally be neglected because of the difficulty in achieving phase matching, $\hat{P}^{(3)}(\vec{r},t)$ can be expressed as follows.

$$\hat{P}^{(3)}(\vec{r},t) = \frac{1}{4} \varepsilon_0 \chi^{(3)} \left| E(\vec{r},t) \right|^2 [2\hat{\sigma} + (\hat{\sigma} \cdot \hat{\sigma}) \hat{\sigma}^*] E(\vec{r},t) \tag{17}$$

If the $E(\vec{r},t)$ is assumed to be linearly polarized along the x-axis, Eq. (17) becomes

$$\hat{P}^{(3)}(\vec{r},t) = \frac{3}{4} \varepsilon_0 \chi^{(3)} \left| E(\vec{r},t) \right|^2 E(\vec{r},t) \hat{x} = \varepsilon_0 \varepsilon_{NL} E(\vec{r},t) \hat{x} \tag{18}$$

where $\hat{x}$ is the x-axis polarization unit vector of the light. ε_{NL} is the nonlinear part of the dielectric constant defined by

$$\varepsilon_{NL} = \frac{3}{4} \chi^{(3)} \left| E(\vec{r},t) \right|^2 \tag{19}$$

The wave equation of Eq. (9) can then be converted into the following form in the frequency domain.

$$\nabla^2 \widetilde{E} + \varepsilon(\omega) k_0 \widetilde{E} = 0 \tag{20}$$

where $k_0 = \omega / c$ and $\varepsilon(\omega)$ is the frequency dependent dielectric constant. The Fourier transform is defined by

$$\widetilde{E}(\vec{r},\omega-\omega_0)=\int_{-\infty}^{\infty}E(\vec{r},t)\exp[i(\omega-\omega_0)t]dt \tag{21}$$

and the frequency dependent dielectric constant is given by

$$\varepsilon(\omega)=1+\widetilde{\chi}^{(1)}(\omega)+\varepsilon_{NL} \tag{22}$$

where $\widetilde{\chi}^{(1)}(\omega)$ is the Fourier transform of $\chi^{(1)}(t)$. Using both the dielectric constant given by Eq. (22) and the definition of $\varepsilon=(n+i\,{}^{\alpha}\!/_{2k_0})^2$ the refractive index n as well as the absorption coefficient α can be found. The intensity dependence of n due to ε_{NL} can be expressed as

$$n=n_0+n_2|E|^2, \tag{23}$$

Using Eqs. (19), (22), and (23), the nonlinear refractive index coefficient n_2 can thus be expressed in the following form after a Taylor series expansion.

$$n_2=\frac{3}{8n_0}\mathrm{Re}(\chi^{(3)}) \tag{24}$$

The temporal phase of an optical signal propagating through optical fiber thus varies as given by the following equation.

$$\phi^{NL}(t)=\frac{2\pi L_{eff}n_2|E(t)|^2}{\lambda} \tag{25}$$

where λ is the signal wavelength. L_{eff} is the effective length of the optical fiber which is defined by $L_{eff} = \dfrac{1-\exp(-\alpha L)}{\alpha}$ where L is the actual fiber length. The time dependent phase change experienced by an optical signal may alternatively be regarded as a change in the instantaneous frequency of the signal, which is referred to as frequency chirp. The SPM induced spectral broadening is thus a consequence of the time dependence of the induced nonlinear phase changes. The generation of new spectral components if coupled with normal dispersion in the fiber will lead to increased temporal broadening of the signal. By contrast in the anomalous dispersion regime signal compression can be achieved, and with the correct conditions, stable pulse forms, which propagate without temporal broadening, can be generated. These special pulses, which have a well-defined pulse shape, are called fundamental solitons.

The principle of SPM in optical fibers has been applied within a range of devices such as the nonlinear optical loop mirror [7], Mach-Zehnder interferometer [21], Kerr nonlinear gate [22] etc.

2.3. Cross-Phase Modulation

When two optical waves are coupled into an opical fiber and propagate through it, the incident waves interact each other through the fiber Kerr nonlinearity. Each waves experiences SPM and they also influence each other by changing the refractive index experienced by the other wave, resulting in phase modulation of the other wave. This phenomenon is known as cross-phase modulation (XPM). The nonlinear refractive index for each wave of a different frequency with the same polarization in an optical fiber can thus be expressed as [19]

$$n_i = n_i^L + n_2 I_i + 2n_2 I_j \tag{26}$$

$$n_j = n_j^L + n_2 I_j + 2n_2 I_i \tag{27}$$

where $n_{i,j}^L$ represent the linear refractive index for each signal and n_2 is the nonlinear refractive index. $I_{i,j}$ represents the optical intensities of the two waves, respectively. The factor of 2 on the right hand side of Eq. (26) and (27) shows that XPM is twice as effective as SPM for the same intensity.

In order to clearly understand the origin of XPM, electromagnetic theory can also be adopted as done in the previous section. The rapidly varying part of two electric fields can be written in the form

$$\begin{aligned}\vec{E}(\vec{r},t) &= \vec{E}_1(\vec{r},t)+\vec{E}_2(\vec{r},t) = \mathrm{Re}[\hat{E}_1(\vec{r},t)e^{-i\omega_1 t}+\hat{E}_2(\vec{r},t)e^{-i\omega_2 t}] \\ &= \mathrm{Re}[\hat{\sigma}_1 E_1(\vec{r},t)e^{-i\omega_1 t}+\hat{\sigma}_2 E_2(\vec{r},t)e^{-i\omega_2 t}]\end{aligned} \tag{28}$$

where $\hat{\sigma}_{1,2}$ is the polarization unit vector, ω_1 and ω_2 are the center frequencies of the two waves. The amplitudes of $E_1(\vec{r},t)$ and $E_2(\vec{r},t)$ are assumed to be slowly varying functions of time. By substituting Eq. (28) in Eq. (16), nonlinear polarization vectors of $\hat{P}^{(3)}_{\omega_1}(\vec{r},t)$ and $\hat{P}^{(3)}_{\omega_2}(\vec{r},t)$ at frequencies of ω_1 and ω_2, each can be obtained.

$$\begin{aligned}\hat{P}^{(3)}_{\omega_1}(\vec{r},t) &= \frac{1}{4}\varepsilon_0\chi^{(3)}\{|E_1(\vec{r},t)|^2 E_1(\vec{r},t)[2\hat{\sigma}_1+(\hat{\sigma}_1\cdot\hat{\sigma}_1)\hat{\sigma}_1^{*}] \\ &+2|E_2(\vec{r},t)|^2 E_1(\vec{r},t)[\hat{\sigma}_1+(\hat{\sigma}_1\cdot\hat{\sigma}_1^{*})\hat{\sigma}_2+(\hat{\sigma}_1\cdot\hat{\sigma}_2)\hat{\sigma}_2^{*}]\}\end{aligned} \tag{29}$$

$$\begin{aligned}\hat{P}^{(3)}_{\omega_2}(\vec{r},t) &= \frac{1}{4}\varepsilon_0\chi^{(3)}\{|E_2(\vec{r},t)|^2 E_2(\vec{r},t)[2\hat{\sigma}_2+(\hat{\sigma}_2\cdot\hat{\sigma}_2)\hat{\sigma}_2^{*}] \\ &+2|E_1(\vec{r},t)|^2 E_2(\vec{r},t)[\hat{\sigma}_2+(\hat{\sigma}_1\cdot\hat{\sigma}_2^{*})\hat{\sigma}_1+(\hat{\sigma}_1\cdot\hat{\sigma}_2)\hat{\sigma}_1^{*}]\}\end{aligned} \tag{30}$$

Higher-order terms are neglected because of the difficulty in achieving phase matching. Similar to Eq. (18), $\hat{P}^{(3)}_{\omega_j}(\vec{r},t)$ can be written in the following form (*j = 1, 2*) if the two waves are linearly polarized at the x-axis [20].

$$\hat{P}^{(3)}_{\omega_j}(\vec{r},t) = \varepsilon_0\varepsilon_j^{NL}\hat{x}E_j \tag{31}$$

Similar to Eq. (22), the frequency dependent dielectric constant can be decomposed into the linear ($\varepsilon_j(\omega)^L$) and nonlinear ($\varepsilon_j(\omega)^{NL}$) parts as follows

$$\varepsilon_j(\omega) = \varepsilon_j(\omega)^L + \varepsilon_j(\omega)^{NL} \tag{32}$$

If the imaginary part of $\chi^{(3)}$ related to the absorption coefficient is ignored.

$$\varepsilon_j = \varepsilon_j^L + \varepsilon_j^{NL} = (n_j^L + \Delta n_j)^2 \tag{33}$$

where n_j^L is the linear part of the refractive index and Δn_j is the change due to the third-order nonlinear effects. Under the approximation of $\Delta n_j << n_j^L = n_0$, the nonlinear part of the refractive index becomes

$$\Delta n_j \cong \varepsilon_j^{NL} \big/ 2n_j \cong n_2(|E_j|^2 + 2|E_{3-j}|^2) \tag{34}$$

where the nonlinear index coefficient n_2 is given by Eq. (24).

As the waves propagate through the fiber, it obtains an intensity-dependent nonlinear phase given by

$$\phi_j^{NL}(t) = \frac{2\pi L_{eff} n_2}{\lambda}[|E_j|^2 + 2|E_{3-j}|^2] \tag{35}$$

The first term accounts for SPM and the second is responsible for XPM. The XPM-induced coupling is also observed between optical waves of the same frequency but different polarization. Interestingly, XPM between the two same frequency waves of orthogonal polarization is less effective than that between two same polarization waves of different frequencies. The XPM effect in an optical fiber has a range of applications, for example a Kerr shutter based on XPM induced nonlinear birefringence [23], wavelength conversion using a dual wavelength nonlinear optical loop mirror [2], optical data demultiplexing [4], etc.

2.4. Four-Wave Mixing

Four-wave mixing (FWM) occurs when two or more frequencies of light waves propagate through an optical fiber. Provided that a condition known as phase matching (momentum conservation) is satisfied, light is generated at new frequencies by converting optical power at the original signal frequencies. The equation that defines the frequency of a FWM product is given by [19]

$$\omega_4 = \omega_1 + \omega_2 - \omega_3 \tag{36}$$

The generation of new frequencies from non-degenerate input signals is shown schematically in Figure 1.

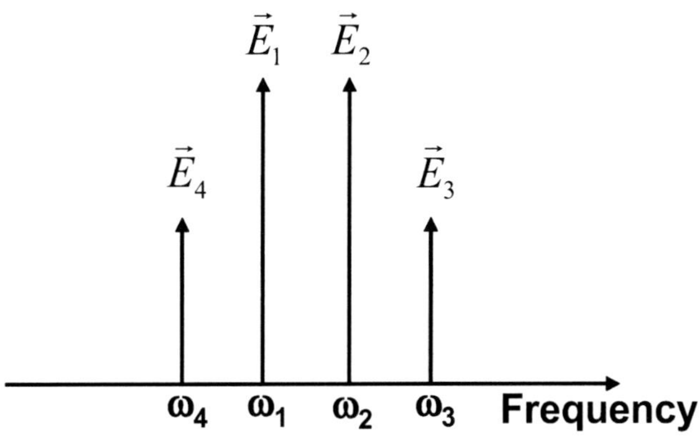

Figure 1. Schematic diagram of additional frequency generation through FWM for the non-degenerate case.

The main features of FWM can also be understood by taking into account the third-order polarization term in Eq. (16). Consider four optical waves at frequencies of ω_1, ω_2, ω_3, and ω_4 which are propagating in the same direction z and assume that their electric fields have the following relation.

$$\left|\vec{E}_1\right|^2, \left|\vec{E}_2\right|^2 >> \left|\vec{E}_3\right|^2, \left|\vec{E}_4\right|^2 \tag{37}$$

The nonlinear polarization vectors of $\hat{P}^{(3)}_{\omega_4}(\vec{r},t)$ at frequencies of ω_4 in the case of symmetric non-degenerate FWM can be expressed as.

$$\hat{P}^{(3)}_{\omega_4}(\vec{r},t) = \frac{1}{2}\varepsilon_0\chi^{(3)}\{|E_1(\vec{r},t)|^2 E_4(\vec{r},t)[\hat{\sigma}_4 + (\hat{\sigma}_1^* \cdot \hat{\sigma}_4)\hat{\sigma}_1 + (\hat{\sigma}_1 \cdot \hat{\sigma}_4)\hat{\sigma}_1^*]$$
$$+ |E_2(\vec{r},t)|^2 E_4(\vec{r},t)[\hat{\sigma}_4 + (\hat{\sigma}_2^* \cdot \hat{\sigma}_4)\hat{\sigma}_2 + (\hat{\sigma}_2 \cdot \hat{\sigma}_4)\hat{\sigma}_2^*]$$
$$+ 2E_1(\vec{r},t)E_2(\vec{r},t)E_3^*(\vec{r},t)[(\hat{\sigma}_1 \cdot \hat{\sigma}_2)\hat{\sigma}_3^* + (\hat{\sigma}_2 \cdot \hat{\sigma}_3^*)\hat{\sigma}_1 + (\hat{\sigma}_1 \cdot \hat{\sigma}_3^*)\hat{\sigma}_2]\} \quad (38)$$

If all the waves are linearly polarized along the x-axis, Eq. (38) becomes

$$\hat{P}^{(3)}_{\omega_4}(\vec{r},t) = \frac{3}{2}\varepsilon_0\chi^{(3)}\hat{x}\{|E_1(\vec{r},t)|^2 E_4(\vec{r},t) + |E_2(\vec{r},t)|^2 E_4(\vec{r},t) + 2E_1(\vec{r},t)E_2(\vec{r},t)E_3^*(\vec{r},t)\} \quad (39)$$

The term proportional to E_4 in Eq. (39) accounts for XPM effects and the remaining term is responsible for FWM. Efficient FWM requires frequency matching as well as phase matching. Two photons at frequencies ω_1 and ω_2 are annihilated with the simultaneous creation of two photons at frequencies ω_3 and ω_4 as described in Eq. (36)

The phase matching condition to be achieved can be expressed as

$$\Delta k = k_3 + k_4 - k_1 - k_2 = {(n_3\omega_3 + n_4\omega_4 - n_1\omega_1 - n_2\omega_2)}/{c} = 0 \quad (40)$$

In the particular case of $\omega_1 = \omega_2$ (degenerate case), it is relatively easy to satisfy $\Delta k = 0$

The generation of new optical components through four-wave mixing has serious negative implications within wavelength division multiplexed (WDM) transmission systems, however it can be used to obtain data demultiplexing [24], wavelength conversion [25], dispersion compensation in long-haul optical fiber links [26], and the like.

2.5. Stimulated Raman Scattering

As well as the electronic response of a nonlinear optical medium to intense optical fields, another sort of nonlinearity related to the interaction between the optical photons and phonons (vibrational state) is also important in optical systems. Since these inelastic nonlinear effects can be regarded as the scattering of a pump beam off a phonon, with the resultant transfer of energy into a lower energy beam, they are called scattering phenomena. First of all, spontaneous scattering transfers a small portion of the input beam power (typically $\sim 10^{-6}$) to another optical beam at a frequency down shifted by an amount determined by the optical phonon vibrational modes of the medium. This phenomenon is called the Raman effect [19]. The incident beam behaves as a pump for generating frequency downshifted light (referred to as the Stokes wave). The spontaneous scattering process is typically a rather weak process that can occur as the result of the medium optical property fluctuations (typically in the dielectric constant) excited mainly by thermal effects. However, when the intensity of the incident beam is large enough to modify the optical properties of the medium, highly efficient light scattering, so called stimulated scattering occurs. Stimulated Raman scattering (SRS) is a very strong scattering process and ten percent or more of the energy of the incident beam is often converted into the Stokes wave. Another difference between spontaneous and stimulated Raman scattering is that the spontaneous process leads to nearly isotropic emission, whilst the stimulated process leads to emission in the forward and backward directions.

The evolution of the Stokes wave and the pump in optical fiber under CW or quasi-CW conditions is governed by the following set of two coupled equations [20],

$$\frac{dI_s}{dz} = g_R I_p I_s - \alpha_s I_s \tag{41}$$

$$\frac{dI_p}{dz} = -\frac{\omega_p}{\omega_s} g_R I_p I_s - \alpha_p I_p \tag{42}$$

where I_s is the Stokes intensity, I_p is the pump intensity, and g_R is the Raman gain coefficient. α_s and α_p represents the fiber loss at the Stokes and pump wavelengths.

Measurements of the Raman gain coefficient g_R for silica fibers were reported in Ref [27]. The Raman gain coefficient g_R depends on the material composition of the fiber core, the waveguide structure, and the pump wavelength.

The peak gain g_R is in conventional standard single mode fiber is known to be ~4.3 x 10^{-14} m/W. Figure 2 shows the normalized Raman gain coefficient g_R for fused silica fiber as a function of frequency shift at a pump wavelength of 1.5 μm, extending from zero frequency shift up to 40 THz [28]. The peak gain is observed at a frequency shift of 13 THz from the pump and the gain has a bandwidth of 10 THz.

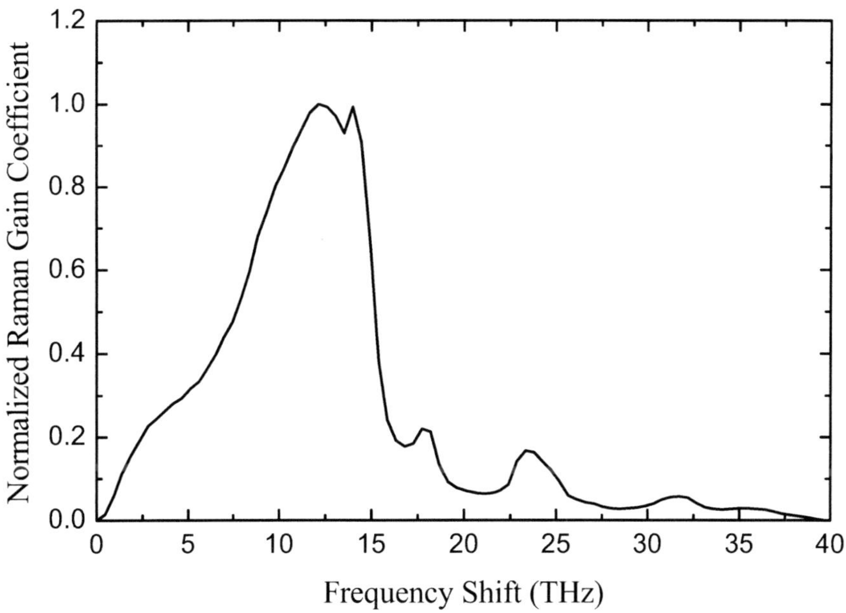

Figure 2. Normalized Raman gain spectrum in fused silica fiber for copolarized pump and signal beams ([28]).

To find the Raman threshold in optical fiber, the coupled equations of Eqs. (41) and (42) need to be solved. Assuming pump depletion is negligible, the solution for I_s is obtained as

$$I_s(L) = I_s(0)\exp(g_R I_0 L_{eff} - \alpha_s L) \quad (43)$$

In order to calculate the Stokes power, an integration over the whole range of the Raman-gain bandwidth should be performed considering each frequency component of energy $\hbar\omega$.

$$P_s(L) = \int_{-\infty}^{\infty} \hbar\omega \exp[g_R(\omega) I_0 L_{eff} - \alpha_s L] d\omega \tag{44}$$

where the fiber is assumed to support a single transverse mode. The Raman threshold is defined as the input pump power, which generates the same amount of Stokes power as residual pump power at the fiber output.

$$P_s(L) = P_p(L) = P_0 \exp(-\alpha_p L) \tag{45}$$

where $P_0 = I_0 A_{eff}$. P_0 and A_{eff} are the input pump power and the effective mode area, respectively. Under the assumption of a Lorentzian Raman-gain spectrum and $\alpha_p = \alpha_s$, the critical pump power is given by [29].

$$g_R P_0^{cr} L_{eff} \Big/ A_{eff} \cong 16 \tag{46}$$

The important applications of the SRS phenomenon in optical fibers include Raman fiber lasers [30] and broadband Raman amplifiers [8]. In particular fiber based Raman amplifiers have attracted huge research attention for ultra-broadband WDM systems applications nowadays since the amplification band can be applied anywhere of the fiber transparency window by simply changing the pump wavelength.

2.6. Stimulated Brillouin Scattering

Stimulated Brillouin scattering is similar to SRS since it also generates a Stokes wave downshifted from the frequency of the incident pump beam. However, the main features of stimulated Brillouin scattering (SBS) are different from those of SRS. For example, the Stokes wave of SBS propagates only in the backward direction in contrast with SRS that occurs in both directions [19]. SBS has a very narrow gain bandwidth of around 10~50 MHz at the frequency shift of ~10 GHz from the pump, which is much smaller than that of SRS. The threshold pump power for SBS depends on the spectral bandwidth of the pump light. Basically SBS is associated with scattering from acoustic phonons whilst optical phonons participate in SRS. The phase matching condition for the SBS process is thus much more stringent due to the nature of the acoustic phonon energy momentum relationship, as compared to the SRS process. Acoustic waves generated through the process of electrostriction by a pump wave cause a periodic modulation of the refractive index in an optical fiber. The index grating scatters the pump wave through Bragg diffraction and the scattered light thus propagates backward relative to the pump wave at a downshifted frequency due to the Doppler shift associated with the motion of the grating, which moves at the acoustic velocity.

For the case of scattering at an angle θ, the Stokes scattering can be illustrated as the diagram in Figure 3. The frequencies and the wave vectors of the related three wave components can thus be expressed by [20]

$$\Omega = \omega_p - \omega_s \tag{47}$$

$$\vec{q} = \vec{k}_p - \vec{k}_s \tag{48}$$

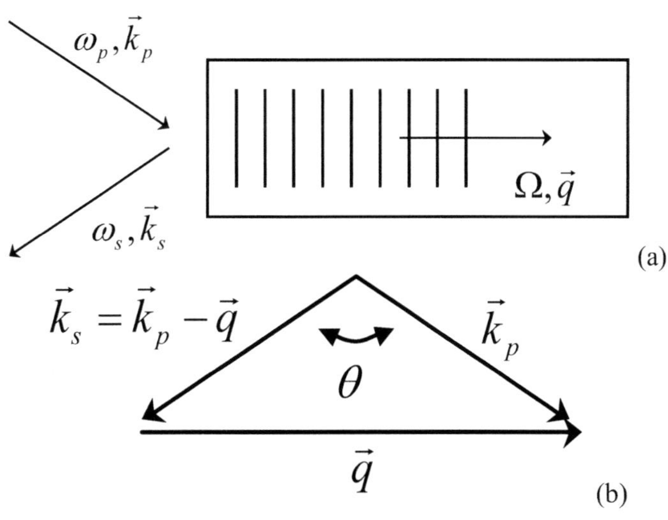

Figure 3. Illustration of the Stokes scattering based on Brillouin effect.

where Ω, ω_p and ω_s are the frequencies, and $\vec{q}$, $\vec{k}_p$ and $\vec{k}_s$ are the wave vectors of the acoustic, pump and Stokes waves, respectively. Since $|\vec{k}_p|$ is almost equal to $|\vec{k}_s|$ (because Ω is much smaller than ω_p), the frequency Ω and the wave vector $\vec{q}$ of the acoustic wave satisfy the dispersion relation [20]

$$\Omega = |\vec{q}|v_A = 2v_A|\vec{k}_p|\sin\left(\theta/2\right) \tag{49}$$

where v_A is the acoustic wave velocity.

Eq. (49) shows that the frequency shift of the Stokes wave depends on the scattering angle. Note that the Stokes shift Ω is equal to zero for forward scattering and is maximum for backscattering ($\theta = \pi$). Eq. (49) predicts that the Brillouin scattering in optical fiber should thus occur only in the backward direction although small amount of spontaneous or thermal Brillouin scattering can occur in the forward direction due to the guide nature of acoustic waves [31]. The Brillouin frequency shift in the backward direction can be expressed as [20]

$$v_B = \frac{\Omega}{2\pi} = \frac{2nv_A}{\lambda_p} \tag{50}$$

where n and λ_p are the refractive index and the pump wavelength, respectively.

In contrast to the case of SRS, the spectral width Δv_B of the Brillouin gain spectrum is very small (~10 MHz). The spectral bandwidth is associated with the damping time of acoustic waves, that is the phonon lifetime T_B. Assuming that an acoustic wave decays as $\exp\left(-t/T_B\right)$, the Brillouin gain coefficient will exhibit a Lorentzian spectral profile as follows.

$$g_B(v) = \frac{\left(\Delta v_B/2\right)^2}{(v - v_B)^2 + \left(\Delta v_B/2\right)^2} g_B(v_B) \tag{51}$$

where $\Delta\nu_B$ is 3dB spectral bandwidth and can be expressed as $\Delta\nu_B = 1/\pi T_B$. The peak value of the Brillouin gain coefficient at $\nu = \nu_B$ is given by [32]

$$g_B(\nu_B) = \frac{2\pi n^7 p_{12}^2}{c\lambda_p \rho_0 \nu_A \Delta\nu_B} \tag{52}$$

where p_{12} is the longitudinal electro-optic coefficient, ρ_0 is the material density, and λ_p is the pump wavelength.

In order to find the SBS threshold, the mutual interaction between the pump and Stokes waves governed by the coupled-intensity equations needs to be solved. However Unlike the SRS case, the sign of dI_s/dz should be changed to account for the counter-propagating nature of the Stokes wave relative to the pump wave. Note that the fiber loss is nearly same for the pump and Stokes waves, i.e. $\alpha_p \cong \alpha_s = \alpha$ due to the relatively small value of the Brillouin shift, $\omega_p \cong \omega_s$. Assuming the pump and Stokes waves are linearly polarized along the same direction and maintain their polarization along the fiber, the coupled equations can be expressed as [20]

$$\frac{dI_s}{dz} = -g_B I_p I_s + \alpha I_s \tag{53}$$

$$\frac{dI_p}{dz} = -g_B I_p I_s - \alpha I_p \tag{54}$$

If pump depletion is neglected, the Stokes intensity is found to be given by

$$I_s(0) = I_s(L)\exp\left(g_B P_0 L_{eff} / A_{eff} - \alpha L \right) \tag{55}$$

With a similar approach to that of Section 2.5, the Brillouin threshold, to a good approximation, is given by [28]

$$g_B P_0^{cr} L_{eff} \Big/ A_{eff} \cong 21 \tag{56}$$

The numerical factor 21 can also change by a factor between 1 and 2 depending on the polarization status of pump and Stokes waves. SBS is unlikely to find applications for signal amplification in telecommunication systems because of its narrow gain bandwidth. However, SBS based fiber lasers have been successfully demonstrated [33].

Chapter 3

KERR NONLINEARITY FIGURE-OF-MERIT IN OPTICAL FIBER

A huge number of researches have been performed to develop optical fibers with high Kerr optical nonlinearity per unit length in that compactness and stability of optical fiber-based nonlinear signal processing devices can be improved by development of such highly nonlinear fibers. Recently, the nonlinear optical fiber fabrication technologies has advanced significantly and a variety of highly-nonlinear fibers are commercially available: for example, highly-nonlinear dispersion-shifted fiber (HNL-DSF) [34], silica-glass-based holey fiber (HF) with small core [35], Bi_2O_3-glass-based nonlinear fiber (Bi-NLF) [15], SF57- based HF [36], and so on.

Optical fiber Kerr nonlinearity parameter γ is defined as follows.

$$\gamma = \frac{2\pi n_2}{\lambda A_{eff}} \tag{57}$$

where n_2 is the material nonlinear refractive index, λ is the signal wavelength, and A_{eff} represents the effective core area of an optical fiber. Optical fibers with high Kerr nonlinearity can thus be fabricated by using novel glass materials with a high nonlinear refractive index (n_2) [15, 36] or/and tightly confining light within the core with holey fiber (or called photonic crystal fiber) structures [35].

One interesting and important issue in design and application of those nonlinear fibers is how to properly define the figure-of-merit (FOM) for the nonlinearity efficiency evaluation. So far, the commonly used, conventional FOMs are $\gamma \cdot L_{eff}$ (L_{eff} is the effective length) [37] and γ/α (α is the fiber propagation loss) [37]. Both of the FOMs are expressed in the unit of W^{-1}. The former provides the quantitative information of the nonlinear phase shift per unit input optical power achievable in an optical fiber when an input pump power is given and the fiber length L is comparable with or longer than L_{eff}. On the other hand, the latter dictates the Kerr nonlinearity value relative to the optical fiber propagation loss. Furthermore, the latter FOM can be considered as a modified form of the former, which is obtainable in the case that the optical fiber length is assumed to be infinite. Note that L_{eff} approaches $1/\alpha$ when the fiber length is extremely long. One noticeable point is that both $\gamma \cdot L_{eff}$ and γ/α cannot be a correct measure for the fiber nonlinearity evaluation since they do not take into account the pump beam power loss induced by SBS, SRS or both. Since the SBS threshold is much lower than that of SRS in general, the SBS effect decides the upper limit of the input pump power available within the fiber.

Therefore, the pump beam power limit caused by the SBS threshold (P_{SBS}) must be taken into account when defining a FOM. In particular, this is true in the case of using pure continuous wave (CW) pump beams: for example, FWM based wavelength conversion [35] and parametric amplification [38]. A more reasonable FOM ($F_{nl\text{-}SBS}$) for the Kerr nonlinearity efficiency evaluation is thus given as [39]

$$F_{nl-SBS} = \gamma \cdot L_{eff} \cdot P_{SBS} \tag{58}$$

which means the maximum nonlinear phase shift available from the nonlinear fiber before SBS induces pump power loss. Table 1 summarizes the definition and the description of the three optical fiber Kerr nonlinearity FOMs.

In order to clearly understand the meaning of the $F_{nl\text{-}SBS}$, let us take four types of state-of-the-art highly nonlinear optical fiber, experimentally determine the three FOMs for the four fiber samples from measurements of γ and P_{SBS}, and compare them one another. The four types of fiber sample are as follows: a 1 m long Bi-NLF, a 15 m long HF, a 150 m long HNL-DSF, and a 1 km long DSF. Fiber lengths L were chosen so that $L \approx L_{eff}$. The detailed parameters such as fiber background loss, group velocity dispersion (GVD), effective length (L_{eff}), and effective core area (A_{eff}) for the four types of fiber are shown in the Table 2.

Table 1. Three Kinds of Optical Fiber Kerr Nonlinearity Figure-of-Merits (FOMs)

FOM	UNIT	DESCRIPTION
γ·Leff	W^{-1}	Nonlinear phase shift per unit input optical pump power achievable in an optical fiber when the fiber length L is comparable with or longer than L_{eff}. No inclusion of input pump power limit by nonlinear scattering effects.
γ/α	W^{-1}	Nonlinear phase shift per unit input optical power achievable in an optical fiber in the case of fiber length L = infinite. No inclusion of input pump power limit by nonlinear scattering effects.
$\gamma \cdot L_{eff} \cdot P_{SBS}$	No unit	Maximum nonlinear phase shift available from the nonlinear fiber before occurrence of stimulated Brillouin scattering causedpump power loss

Table 2. Basic Fiber Parameters (after Ref. [39])

Parameters	1 m Bi-NLF	15 m HF	150 m HNLF	1 km DSF
Loss (dB/m)	0.8	0.012	0.0016	0.0003
GVD (ps/nm-km) @ 1550nm	-260	38	1.7	-1.6
L_{eff} (m)	0.98	14.7	146	985
A_{eff} (μm²)	3.08	6.4	11	44

Figure 4 shows measurement setups for the Kerr nonlinearity parameter (a) and the Brillouin gain spectrum (b) [40]. The nonlinearity measurement method was based on the direct CW measurement of the nonlinear phase shift suffered by a beat signal propagating in optical fiber [41]. Output signals from two tunable lasers with a relative wavelength offset of 0.5 nm are first combined using a 50:50 coupler to generate a beat signal and amplified with an erbium-doped fiber amplifier (EDFA) to obtain an adequate power level to generate the FWM effect inside the fiber. The relative intensity ratio between the signal peak and the generated first side band peak was then measured with an optical spectrum analyzer while the launched beat signal power was changed. The relative intensity ratio between the signal peak and the first side band peak gives the nonlinear phase shift through the following equation [41].

$$\frac{I_0}{I_1} = \frac{J_0^2(\phi_{SPM}/2) + J_1^2(\phi_{SPM}/2)}{J_1^2(\phi_{SPM}/2) + J_2^2(\phi_{SPM}/2)} \tag{59}$$

where $I_{0,1}$ are the intensities of the signal peak and the first side lobe peak, respectively, and J_n is the nth Bessel function. ϕ_{SPM} represents the nonlinear Kerr phase shift expressed as

$$\phi_{SPM} = \frac{2\omega_0 n_2 L_{eff} P}{cA_{eff}} \tag{60}$$

where ω_0 is a beat signal frequency and P is power of the beat signal.

For the Brillouin gain spectrum measurement the modulator-based frequency sweeping method is employed [42]. A 1550-nm beam from a tuanble laser was first split into two separate channels using a 50:50 coupler. The first channel was amplified and the background amplified-spontaneous emission was filtered out by using a bandpass filter. The amplified beam was then launched into the fibers under test via a circulator as a SBS pump light. The other channel was intensity-modulated at a frequency of f_s and then amplified with an EDFA. The higher frequency sideband component of the modulated beam was filtered out by a high-precision, fiber Bragg grating (FBG) filter, while the residual carrier and lower sideband components were backwardly coupled into the fibers under test. The lower sideband component served as a SBS probe light. By sweeping the frequency of f_s and detecting the beat signal of the carrier and the amplified lower sideband component with a network analyzer, it was possible to measure the Brillouin gain spectrum.

Figure 5 shows the measured nonlinear phase shift per unit length for each of the sample fibers. The highest nonlinear phase shift per unit length was observed in the Bi-NLF. The estimated nonlinearity parameters γ for the Bi-NLF, the HF, the HNL-DSF, and the DSF were 1100, 56.2, 15.5, and 2.4 $km^{-1}\cdot W^{-1}$, respectively. Using the measured nonlinearity parameters γ the FOMs of $\gamma\cdot L_{eff}$ were calculated for each of the fibers and those were 1.077, 0.826, 2.274, and 2.386 W^{-1} for the Bi-NLF, the HF, the HNL-DSF, and the DSF, each. From a viewpoint of this conventional FOM concept without considering the SBS-caused pump power limit, the DSF should be the best Kerr nonlinearity efficiency. Furthermore, the Bi-NLF and the HF would not be nonlinear media as efficient as the DSF or the HNL-DSF.

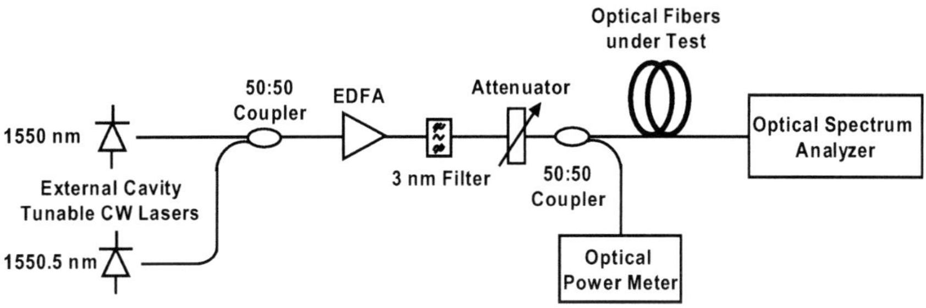

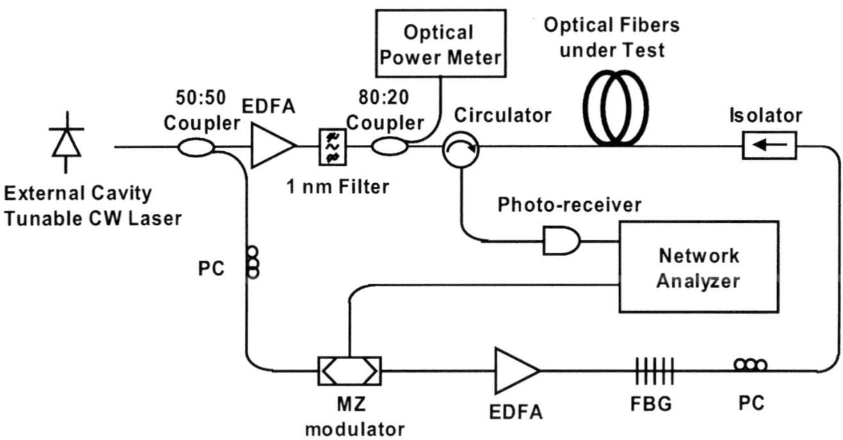

Figure 4. Measurement setups for Kerr nonlinearity and Brillouin gain spectrum (after Ref. [40] © 2006 IEEE).

Then, the FOMs of γ/α were calculated for each fiber and those were 5.97, 20.34, 42.08, and 34.74 W^{-1}, respectively. Interestingly, This FOM definition dictates that the 150-m HNL-DSF would be the best fiber in terms of Kerr nonlinearity, whereas the 1-m Bi-NLF is still considered to be the worst choice although it exhibits the largest nonlinear phase shift per unit length for a given pump beam power. It must be noticed that both of the conventional FOMs neglect a critical parameter for fair Kerr nonlinearity evaluation that is the maximum pump beam power that can propagate through the whole length of the fiber.

Physically, the maximum pump beam power is determined by the SBS threshold in the case of pure CW pump beams, since the SBS threshold is lower than the SRS threshold for pure CW beams. However, some cases like phase modulation-induced spectrum-broadened beams and ultra-short pulse beams exhibit lower SRS thresholds than SBS. Therefore, it can be easily expected that the inclusion of the SBS-induced pump power limit should provide totally different results from those by the conventional FOMs.

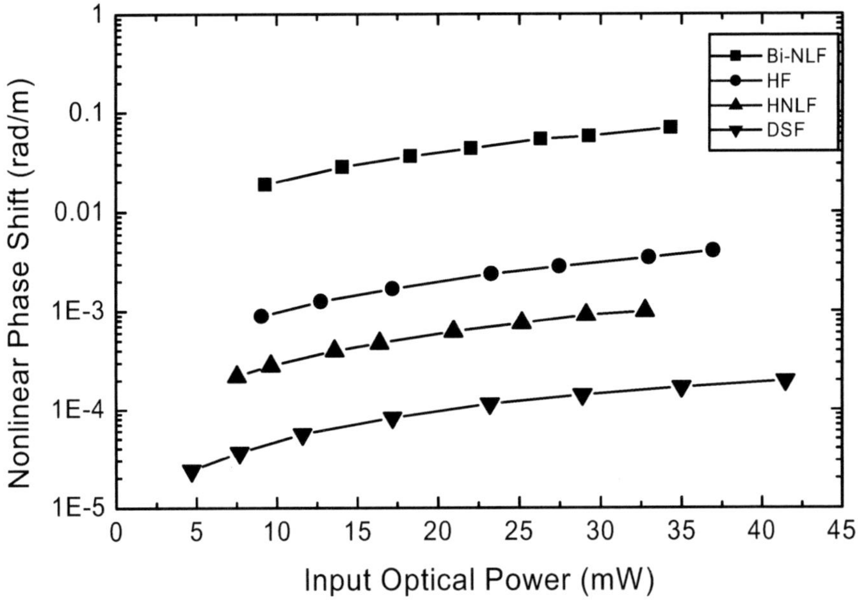

Figure 5. Nonlinear phase shift per unit length measured as a function of the input pump power (after Ref. [39]).

In order to calculate the $F_{nl\text{-}SBS}$ for the four fibers, their SBS threshod values are required. Figure 6(a) shows measured Brillouin gain spectra for the four fibers. All of the four fibers are observed to have Lorentzian gain profiles although a small sideband peak appeared in the HF spectrum. Interestingly, the Bi-NLF shows a Stokes frequency shift of 8.81 GHz, which is significantly different from the values of the silica-based HF and the conventional DSF. The HNL-DSF is also observed to have a lower Stokes frequency shift than the silica-based HF and conventional DSF, which can be attributed to the high Ge doping level inside the core [43]. Figure 6 (b) shows the measured Brillouin gain as a function of the input pump power for the four fibers, and the highest Brillouin-gain slope efficiency exists in the curve of the conventional DSF. The Brillouin gain coefficients (g_B) can be calculated from the values shown in Figure 6 (b), and then the corresponding SBS thresholds (P_{SBS}) can be estimated for the four types of fiber using Eq. (56). Estimated SBS thresholds for the Bi-NLF, the HF, the HNL-DSF, and the DSF are1.027, 0.69, 0.22, and 0.152 W, respectively. These SBS threshold values can be interpreted as the highest CW pump powers that can be launched into each of the fibers.

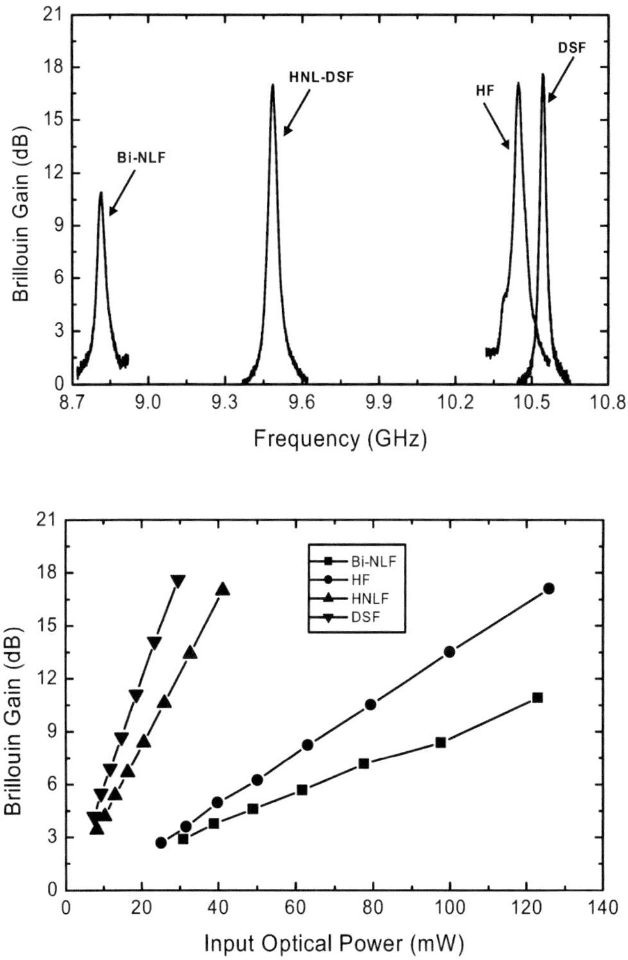

(a)

(b)

Figure 6. (a) Measured Brillouin gain spectra of the four types of fibers at a pump wavelength of 1550 nm. (b) Measured Brillouin peak gain vs input pump power for the four types of fibers (after Ref. [39]).

Table 3. Measured Nonlinearity and Brillouin Parameters (after Ref. [39])

Parameters	1 m Bi-NLF	15 m HF	150 m HNLF	1 km DSF
Nonlinearity γ ($W^{-1}\cdot km^{-1}$)	1100	56.2	15.5	2.4
Nonlinear coefficient n_2, ($m^2\cdot W^{-1}$)	8.17×10^{-19}	8.87×10^{-20}	4.2×10^{-20}	2.6×10^{-20}
Brillouin Frequency Shift (GHz)	8.81	10.44	9.48	10.55
Brillouin Gain Bandwidth (MHz)	32	22.7	20.6	14.5
Brillouin Gain Coefficient g_B ($m\cdot W^{-1}$)	6.43×10^{-11}	1.33×10^{-11}	7.19×10^{-12}	6.17×10^{-12}
SBS Threshold, P_{SBS} (W)	1.027	0.69	0.220	0.152

Table 4. Estimated Nonlinearity Figure-of-Merits

Figure of Merits	1 m Bi-NLF	15 m HF	150 m HNLF	1 km DSF
γL_{eff} (W^{-1})	1.077	0.826	2.274	2.386
γ/α (W^{-1})	5.97	20.34	42.08	34.74
$F_{nl\text{-}SBS}$ ($\gamma L_{eff} \cdot P_{SBS}$)	1.106	0.57	0.50	0.364

Using the measured nonlinear parameters and the SBS thresholds, the values of $F_{nl\text{-}SBS}$ for the Bi-NLF, the HF, the HNL-DSF, and the DSF are calculated to be 1.106, 0.57, 0.50, and 0.364, respectively. All of the measured nonlinearity and Brillouin parameters for the four types of fibers are summarized in Table 3. The estimated FOMs for the four fiber samples are illustrated in Table. 4. It is clearly evident from the calculated values of F_{nl-SBS} that the most efficient nonlinear fiber among the four is the Bi-NLF, and this is a very different result from those by the conventional FOMs. Using Eqs. (56), (57), and (58) $F_{nl\text{-}SBS}$ can also be expressed as

$$F_{nl-SBS} = \frac{42\pi}{\lambda} \cdot \frac{n_2}{g_B} \tag{61}$$

In this equation $F_{nl\text{-}SBS}$ is found to be determined by only the ratio of the nonlinear refractive index (n_2) and the Brillouin gain coefficient (g_B). g_B includes both of the material and waveguide properties in an optical fiber, whilst n_2 is solely determined by the material property. From the perspective of $F_{nl\text{-}SBS}$, it is self-evident that the Bismuth oxide glass is superior to the silica glass for the nonlinear optical fiber fabrication even in the case of conventional core-cladding structures. Note that the HF has a better $F_{nl\text{-}SBS}$ than the HNL-DSF and the conventional DSF although all of the three fibers use silica as a host material. This is believed to be due to the waveguide property discrepancy between the holey structure and the conventional core-cladding structure. The holey fiber structure appears to allow for flexibly controlling the waveguide property to reduce the Brillouin gain coefficient (g_B) compared to the conventional core-cladding fiber structure.

Chapter 4

PHOTONIC CRYSTAL FIBER TECHNOLOGY FOR NONLINEAR SIGNAL PROCESSING DEVICES

4.1. PHOTONIC CRYSTAL FIBER TECHNOLOGY

Photonic crystal fiber (PCF) technology has progressed rapidly in recent years and has resulted in the development of a wide range of optical fibers with unique and highly useful optical properties [44]. PCFs, which are also called microstructured fibers or holey fibers (HFs) possess a core area surrounded by a cladding region composed of a fine array of micrometer-sized air holes that extend along the full fiber length. PCFs are classified into two types depending on light guiding mechanism: index guiding type fiber and photonic bandgap fiber (PBF). In the case of the index guiding type fiber a fiber core is formed by embedding a local solid area of higher refractive index within the photonic microstructure. Figure 7 shows an idealized index guiding type PCF structure. The index guiding type PCFs are typically made of a single material, usually pure silica, and guide light through a modified form of total internal reflection since the volume average index in the core region of the fiber is greater than that of the surrounding microstructured cladding. Note that the hole diameter (d) and pitch Λ, hole to hole spacing), which are the critical design parameters used to specify the structure of an PCF are typically on the scale of the wavelength of light. On the other hand, the light guiding mechanism of PBFs is totally different from that of index guiding type PCFs. PBFs employ photonic bandgap effect to prevent light from escaping from a hollow air core. The lower refractive index of the core than the air-hole cladding cannot support the conventional total internal reflection-based guiding mechanism. In photonic bandgap structures, however, incident light at frequencies where the structure photonic bandgap exists, is trapped and guided along the hollow air core.

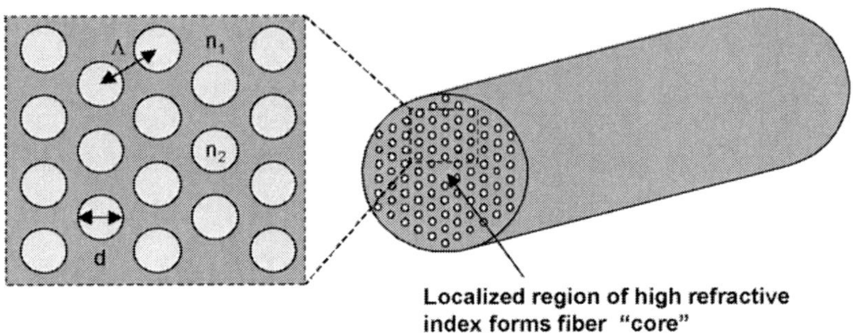

Figure 7. An idealized structure of photonic crystal fiber.

The fundamental physical differences between PCFs and conventional step-index fiber types arise from the way that the guided mode experiences the cladding region. In a conventional fiber, this is largely independent of wavelength to first-order. However, in a PCF, the large index contrast between glass and air and the small structure dimensions combine to make the effective cladding index a strong function of wavelength. Short wavelengths remain tightly confined to the core, and so the effective cladding index is only slightly lower than the core index. However, at longer wavelengths, the mode samples more of the cladding, and so the effective index contrast is larger. This unusual wavelength dependence leads to a host of unique and tailorable optical properties. One striking property is that fibers with a low air fill fraction (d/Λ) can be single-moded regardless of operating wavelength [45]. This property is particularly significant for broadband or short wavelength applications. Tailoring the scale of the cladding features allows the effective fundamental mode area of a holey fiber at 1.55 μm to be varied over three orders of magnitude from ~1 μm^2 to ~1000 μm^2 [46]. PCFs can thus be seen to have a significantly broader range of optical properties than conventional optical fibers, which as well as being of fundamental scientific interest, should also open up the possibility for new and technologically important fiber devices.

The wide single-mode operation of PCF with a low fill fraction ($d/\Lambda < 0.4$) can be understood qualitatively as following. In a standard step-index fiber the number of guided modes is determined by the normalized frequency, or V value [47].

$$V = \frac{2\pi}{\lambda} \cdot \rho \cdot \sqrt{n_{co}^2 - n_{cl}^2} \tag{61}$$

where n_{co} and n_{cl} represent core and cladding indices, respectively. ρ is core radius. The V number should be less than 2.405 for single mode operation in fiber, which means most standard single mode fibers are multi-mode at short wavelengths. For example, SMF28 fiber have a single mode cut-off wavelength of ~1.2 μm. In PCF the light field confines itself in the silica core area rather than distributing itself across the air cladding at short wavelengths, keeping the V number nearly constant and extending the single-mode bandwidth. A more detailed, quantified study of endless single mode operation of PCFs can be found in Ref. [45].

One of the most exciting possibility afforded by PCF technology is the opportunity to develop fibers with a very high optical nonlinearity per unit length [14, 48]. Tailoring the scale of the cladding features allows the effective fundamental mode area of a holey fiber at 1.55 μm to be extremely small (A_{eff} ~ 2 μm^2) [48]. In such a fiber, modest optical powers can induce significant nonlinear effects. The state-of-the-art PCF technology allows the fabrication of fibers with very tightly confined modes, and thus very high optical nonlinearities per unit length. Indeed, a silica holey fiber can have a nonlinearity 10 ~ 100 times higher than that of a conventional silica fiber. Thus, One of the most promising applications of PCF's is in the development of nonlinear optical signal processing devices for fiber-optic communication systems. HF's can have much higher nonlinearity per unit length than conventional fibers, and devices based on such fibers can thus be much shorter in length, and/or operate at lower power levels. Note that further significant increases in fiber nonlinearity should be achievable using fibers made of other glasses, such as the Chalcogenides [49], which have around two orders of magnitude higher nonlinear optical coefficient than silica. In fact, University of Southampton produced the first results in this direction and demonstrated a HF in SF57 lead glass with $\gamma = 640\ W^{-1} \cdot km^{-1}$, about 500 times more nonlinear than conventional SMF28 fiber demonstrated [50].

The unusual wavelength dependence of the effective refractive index in PCF also leads to a range of novel dispersion properties, which are relevant for both linear and nonlinear device applications. For example, silica fibers with a small pitch ($\Lambda < 2$ μm) and large air holes (d) can exhibit anomalous dispersion down to wavelength as low as 550 nm [51]. This has made the generation and propagation of optical solitons in the near-IR and visible regions of the spectrum a reality [52, 53] and is something not possible in conventional single mode fibers. It is also possible to design PCFs with extremely high values of normal dispersion [54] and for example, values as high as -2000 ps/nm-km have been predicted, which suggests that these fibers may find application in dispersion compensation. Other work has shown that broadband dispersion-flattened PCFs can also be designed [55], a property that is likely to be useful for the development of broadband WDM devices: for instance, wavelength converters [34] and parametric amplifiers [56].

4.2. Nonlinear Thresholding Device

The practical advantages of the use of highly nonlinear PCFs in all-optical nonlinear signal processing devices were first experimentally demonstrated in the construction of a 2R regenerative switch for pulse reshaping and data bit amplitude restoration [22]. The 2R regenerative operation can obtained by combining self-phase modulation and offset narrowband spectral filtering [11]. The 2R data regenerative switch was based on a PCF with an effective mode area A_{eff} of just 2.8μm^2 and a nonlinearity $\gamma = 31W^{-1}\cdot km^{-1}$ at 1.55 μm. Similar devices based on conventional fibers are typically of order 1 km in length, whereas in the PCF-based device demonstration [22] just 3.3m of PCF was needed for an operating power of 15W. An 8.7m long variant of this switch was demonstrated to provide an optical thresholding function within an optical code division multiple access (OCDMA) system [57].

A schematic of the superstructured fiber Bragg grating (SSFBG) based OCDMA system incorporating the PCF based nonlinear thresholder is shown in Figure 8. Figure 9(a) shows the spectrum of 2.5 ps soliton pulses both prior and after propagation through the highly nonlinear PCF. As can be seen from the graph new spectral components are generated at both red and blue shifted frequencies relative to the incoming spectrum. Figure 9(b) shows the pulse power transmitted through a 1.0 nm narrowband filter which was offset spectrally by + 2.5 nm relative to the central wavelength of the incident pulses as a function of incident pulse peak power. The S-shaped characteristic is suitable for optical thresholding applications. Such a switch was employed prior to the OCDMA receiver to eliminate the low-level pedestal components arising from the matched filtering of coded bits and obtained a significant improvement in the overall system performance. As shown in Figure 10, the observed ~3 dB power penalty associated with matched filtering alone is totally eliminated.

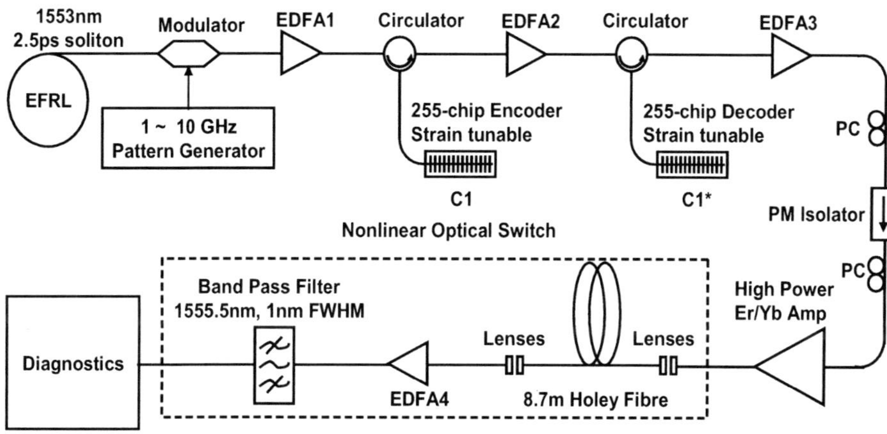

Figure 8. A schematic of the OCDMA system incorporating the PCF based nonlinear thresholder. (after Ref. [57] © 2002 IEEE).

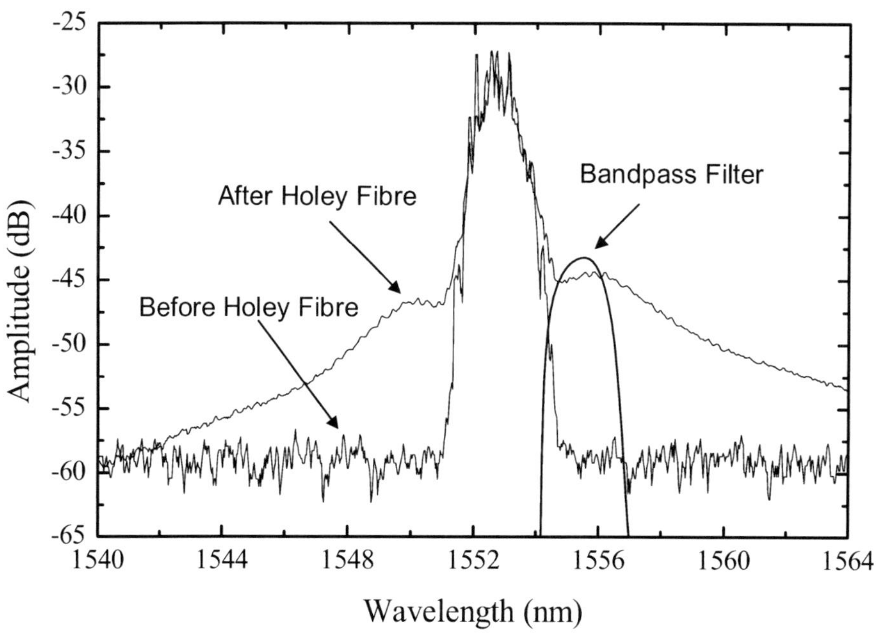

(a)

Figure 9. Continued on next page

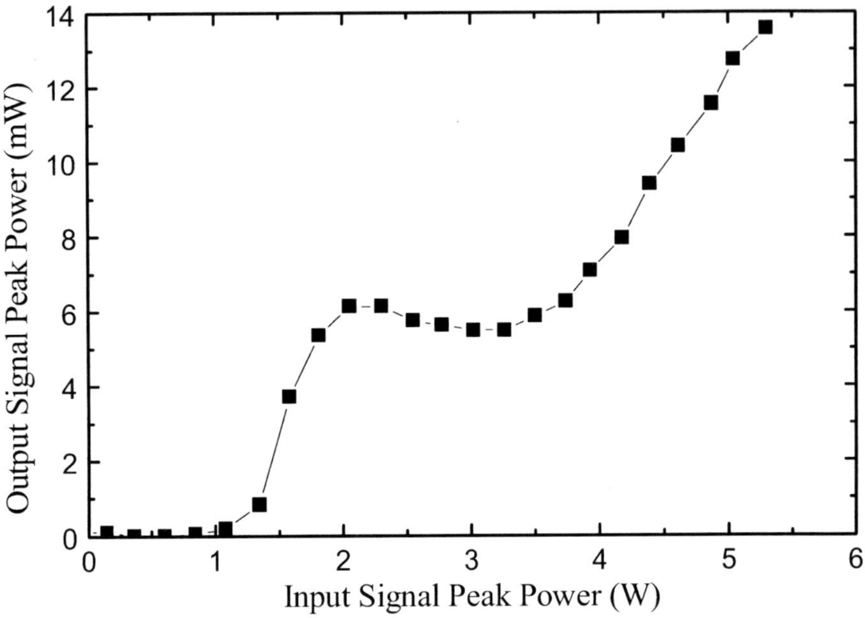

(b)

Figure 9. (a) Spectrum of 2.5 ps soliton pulses both prior and after propagation through the highly nonlinear PCF. (b) Output pulse peak power transmitted through a 1.0 nm narrowband filter which was offset spectrally by + 2.5 nm relative to the central wavelength of the incident pulses as a function of incident pulse peak power. (after Ref. [57] © 2002 IEEE).

4.3. PCF-Based Raman Amplifier

Such fibers also offer reduced length/power requirements for Raman effect based nonlinear devices. Optical devices based on the Raman effect in optical fibers often suffer drawbacks that are not straightforward to overcome. For example long lengths of fiber are required (~10 km) for reasonable pump powers and the related issue of Rayleigh scattering in such fibers compromises their performance [58]. The use of a relatively short length of PCF could probably provide an easy solution to such problems. To obtain an adequate Raman amplification signal gain in a short length of optical fiber specialty fibers with either a very high Raman gain coefficient or a small effective area is needed. PCF, which allows for extremely tightly confined mode is thus a suitable candidate.

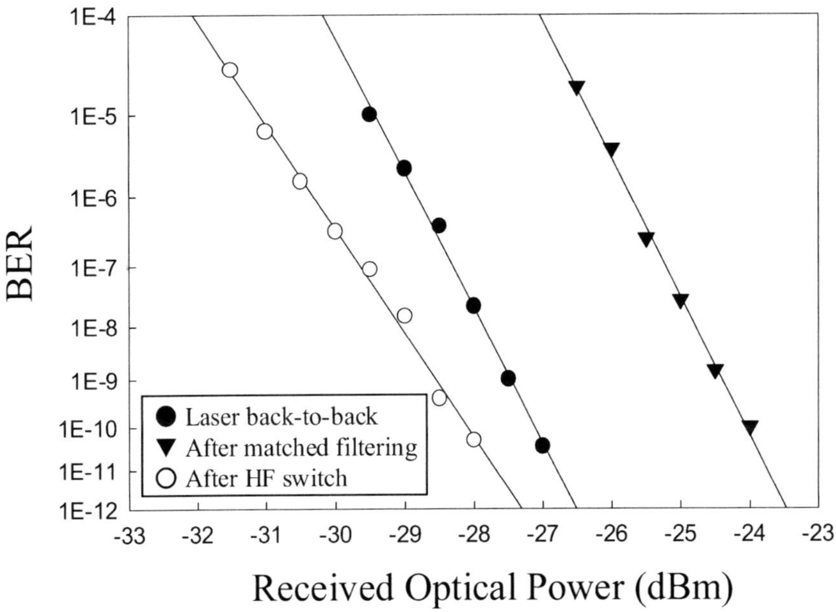

Figure 10. Measured BER versus received optical power at 1.25 Gbit/s. (after Ref. [57] © 2002 IEEE).

A schematic of the PCF based Raman amplifier is shown in Figure 11 [59]. A 75 m long PCF with an effective area of $A_{eff} = 2.85(+/-0.3)\ \mu m^2$ was used for the experiment. The maximum peak power from the pulsed mode pump source coupled into the PCF was ~6.7 W, and the maximum launched signal power ~ -10 dBm. Figure 12(a) shows a measured time-averaged output spectrum from the HF. According to the optical spectrum of background ASE in the graph 1647 nm was found to be the Raman gain peak, which has 13.2 THz frequency separation from the pump wavelength of 1535 nm. Figure 12(b) shows internal Raman gain and noise figure for various probe signal wavelengths and fixed pump/signal powers. Small signal gains as large as 42.8 dB, and noise figures as low as 6 dB were obtained at 1640 nm.

4.4. Raman Modulator

As well as investigating the performance of the PCF as a pure Raman amplifier a Raman modulation/erasure experiment was also carried out. In this instance a strong pump beam at a long wavelength is used to induce loss for a copropagating beam at shorter wavelengths [59, 60]. The same experimental configuration as in the Raman amplification experiment was used other than that the tunable 1600 nm signal source was exchanged for a 20 dBm, 1458 nm CW semiconductor diode laser. In the presence of the pump pulses, the signal beam experiences stimulated Raman scattering (SRS), resulting in an effective nonlinearly induced signal loss. This manifests itself in the time domain through the formation of 'dark' pulses at the signal wavelength where the signal overlaps the pump pulses. Measured oscilloscope traces are shown in Figure 13(a). Modulation extinction ratios in excess of ~11 dB were obtained for peak powers of ~5 W and above as shown in Figure 13(b).

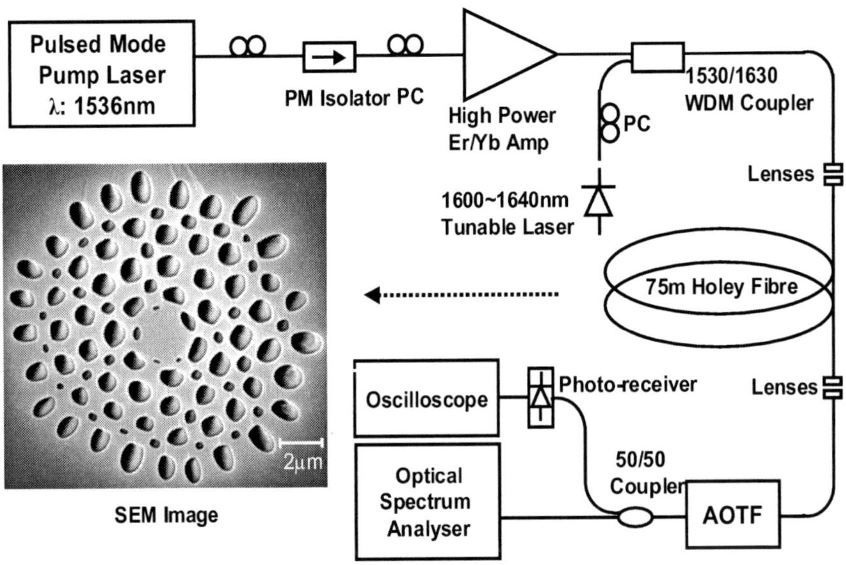

Figure 11. A schematic of the HF based Raman amplifier. (after Ref. [59])

4.5. Tunable Wavelength Converter Based on XPM and Subsequent Spectral Filtering

Wavelength conversion is another important function required within current high-capacity WDM systems. PCF technology was also applied to multiple wavelength conversion over a ~15nm bandwidth at a data rate of 10 Gbit/s using a combination of XPM in a short length of PCF, and narrowband spectral filtering

[61]. The experimental setup is shown in Figure 14. XPM between the control signal and the CW beams results in chirping of the CW laser beam where these beams overlap temporally within the 6m length of the HF.

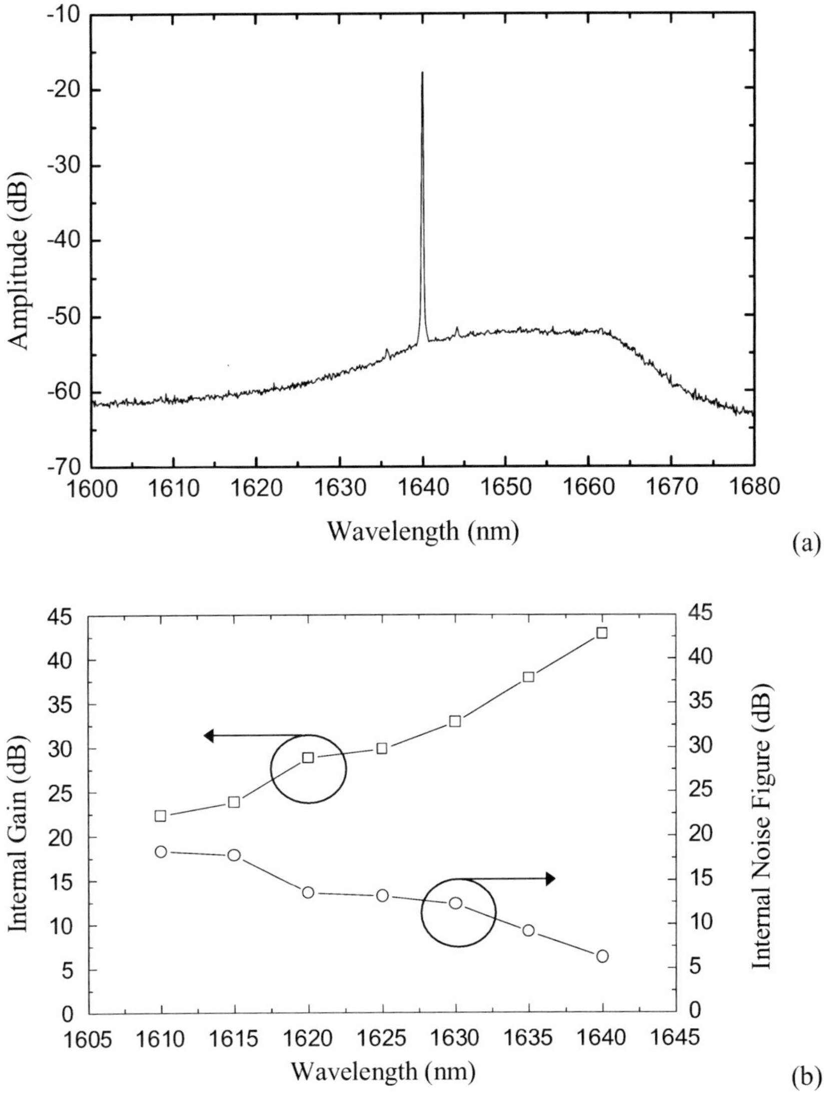

Figure 12. (a) Measured (high gain) amplifier spectrum showing Raman ASE spectrum. (b) Internal Raman gain and noise figure for various probe signal wavelengths (signal power: -10 dBm, pump peak power: 6.7 W). (after Ref. [59]).

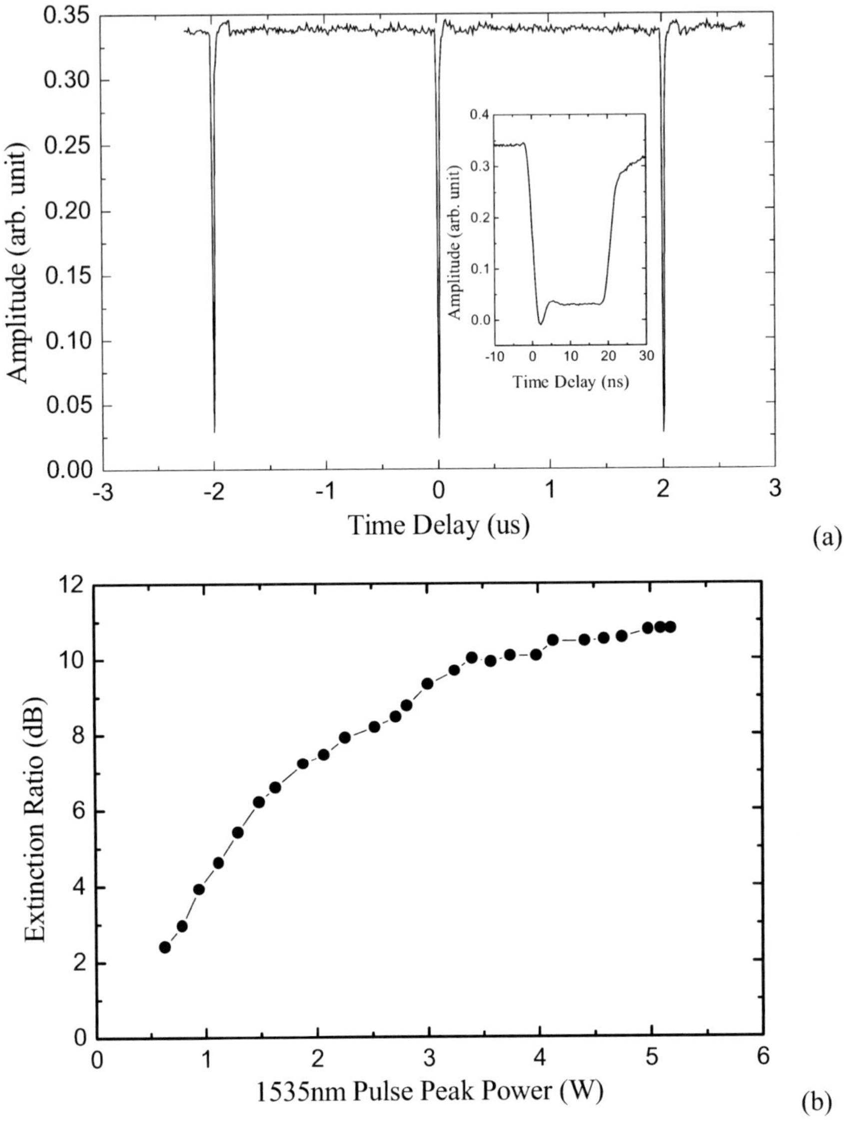

Figure 13. (a) Temporal profile of dark pulses at the SRS modulator output; inset: Close up view of the square-shaped dark pulse (the temporal dip at the falling edge is due to ringing of the photo-receiver). (b) Extinction ratio of SRS based signal modulation versus pump pulse peak power. (after Ref. [59]).

This frequency chirping can then be converted to a frequency converted signal by passing the signal through a narrowband filter which serves to eliminate the residual unchirped CW signal as well as to select one of the two XPM-induced side bands. The measured output spectrum from the PCF is shown in Figure 14. It is clearly evident that each probe beam is spectrally broadened due to XPM induced by the control signal.

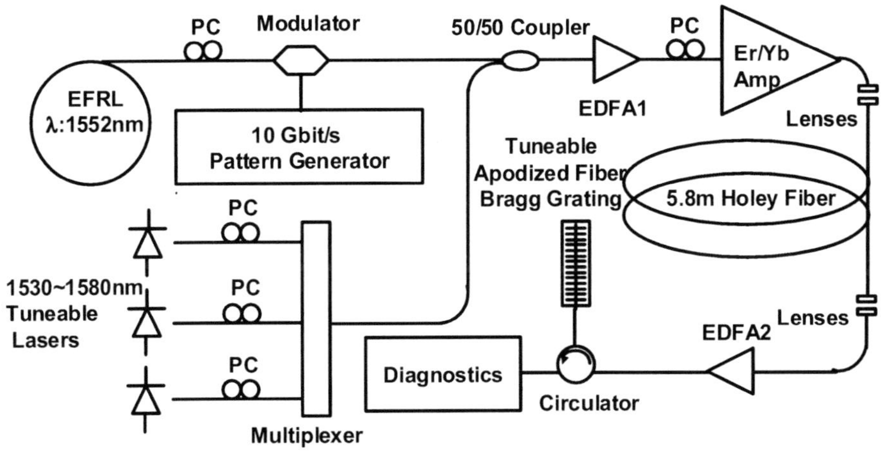

Figure 14. A schematic of a PCF based wavelength converter using the XPM and filtering principle. (after Ref. [61] © 2003 IEEE)

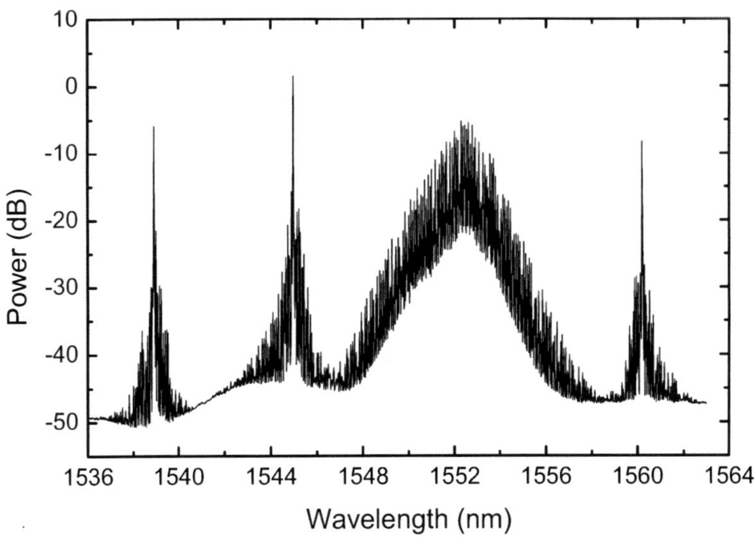

Figure 15. Measured optical spectrum of the three wavelength converted signals and of the control pulses after the 6m HF. (after Ref. [61] © 2003 IEEE)

The pulsewidths of the converted pulses were observed to be almost constant at ~11 ps over a wavelength range of ~15nm Figure 15 shows both the measured BER and eye diagrams for the soliton control and wavelength converted 10 Gbit/s pulses. Error-free, almost penalty-free (~0.5 dB power penalty) wavelength conversion performance was achieved.

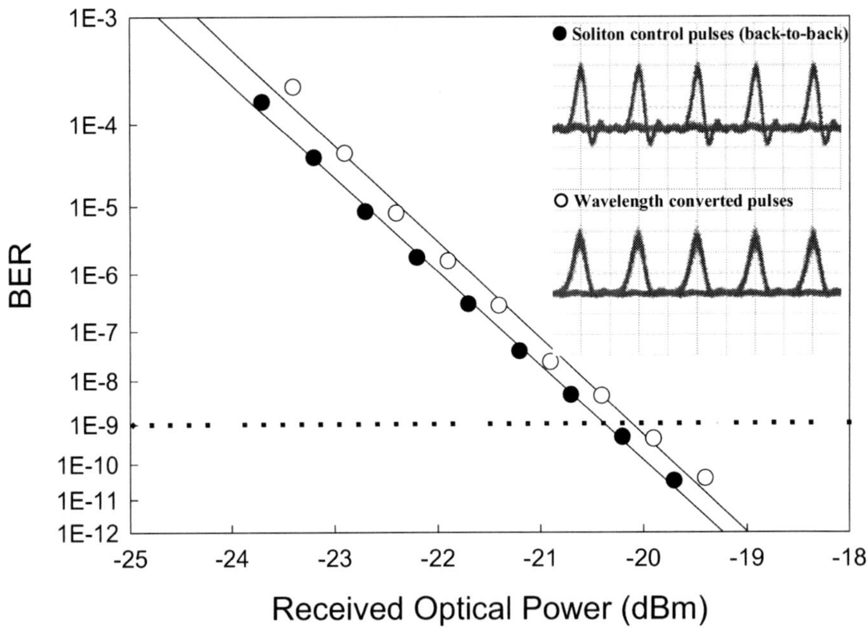

Figure 16. Measured BER versus received optical power for wavelength conversion of 10Gbit/s data pulses to a wavelength of 1545 nm. Inset: Measured eye diagrams. (after Ref. [61] © 2003 IEEE)

4.6. FWM-BASED WAVELENGTH CONVERTER USING A PCF WITH A HIGH SBS THRESHOLD

An alternative, and more flexible wavelength conversion approach, is based on FWM. FWM is particularly attractive for wavelength conversion due to its transparency to both bit-rate and modulation format. Highly efficient, broadband wavelength conversion based on FWM in optical fiber requires high nonlinearity, small dispersion, a low dispersion slope, and a short fiber length to reduce the phase mismatch between the interacting waves [62, 63]. In addition, a high-stimulated Brillouin scattering (SBS) threshold is required to suppress SBS induced pump power loss in order to permit higher conversion efficiency [64]. Thus, highly nonlinear PCF with a high SBS threshold could be a powerful nonlinear medium for FWM based wavelength conversion.

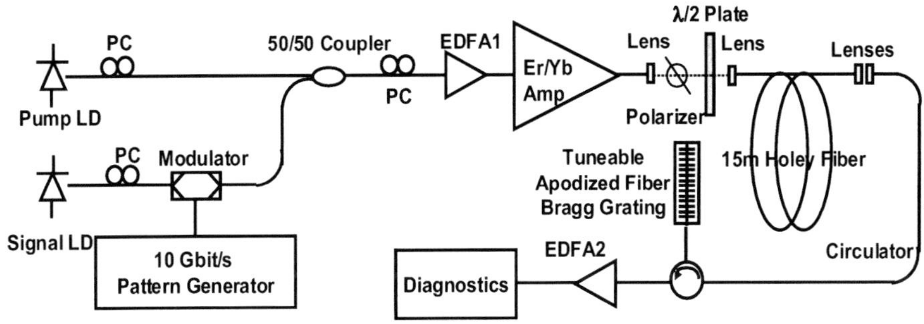

Figure 17. A schematic of a FWM based wavelength converter using a PCF with a high SBS threshold. (after Ref. [35] © 2003 IEEE)

A tunable FWM based conversion using a highly nonlinear PCF with a high SBS threshold was successfully demonstrated [35]. The experimental schematic is shown in Figure 17. A highly nonlinear PCF with structural nonuniformity along its length was used to obtain a high SBS threshold [65]. The SBS threshold of the 15 m HF was measured to be over ~120 mW, around two times higher than would be expected for a silica fiber with a uniform cross-sectional profile. In Figure 18(a) a strong FWM wavelength converted signal is clearly evident at a wavelength of 1544 nm despite the short fiber length used and both second, and third order idler beams are also observable.

Figure 18(b) shows the measured conversion efficiency. A maximum conversion efficiency of –16 dB was achieved over a 3 dB bandwidth of ~10 nm. BER measurements were performed on the wavelength-converted channel at 1544 nm. Figure 19 shows both the measured BER and eye diagrams for both the 10 Gbit/s NRZ input signal and wavelength converted output signal. Error-free wavelength conversion performance was obtained with a ~2 dB power penalty relative to that of the back-to-back input signal.

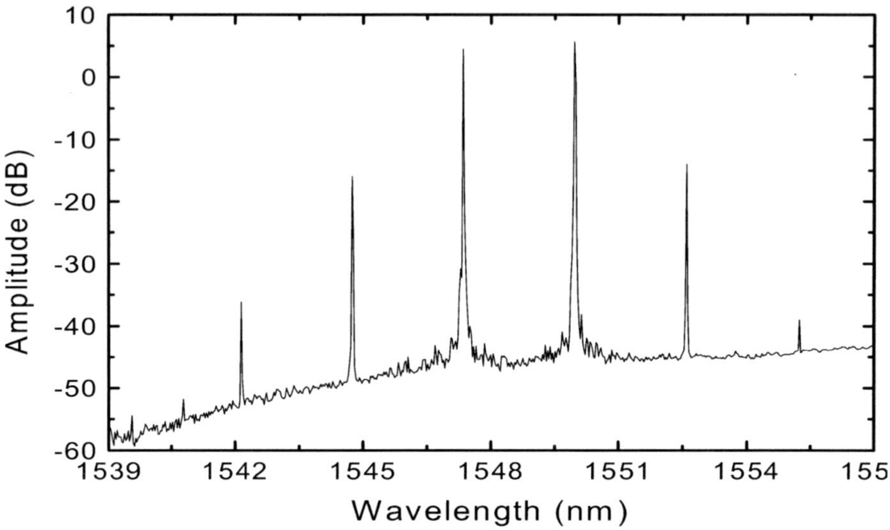

(a)

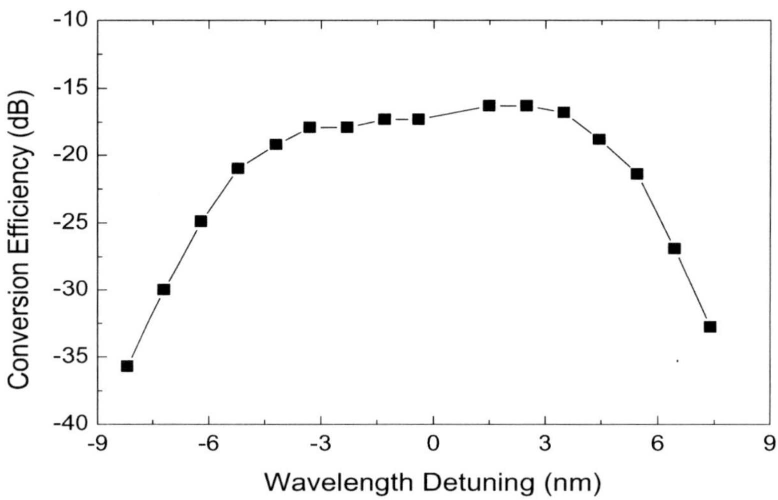

(b)

Figure 18. (a) Output optical spectrum from the HF. (b) Measured conversion efficiency versus wavelength detuning relative to a fixed signal wavelength of 1550 nm. (after Ref. [35] © 2003 IEEE)

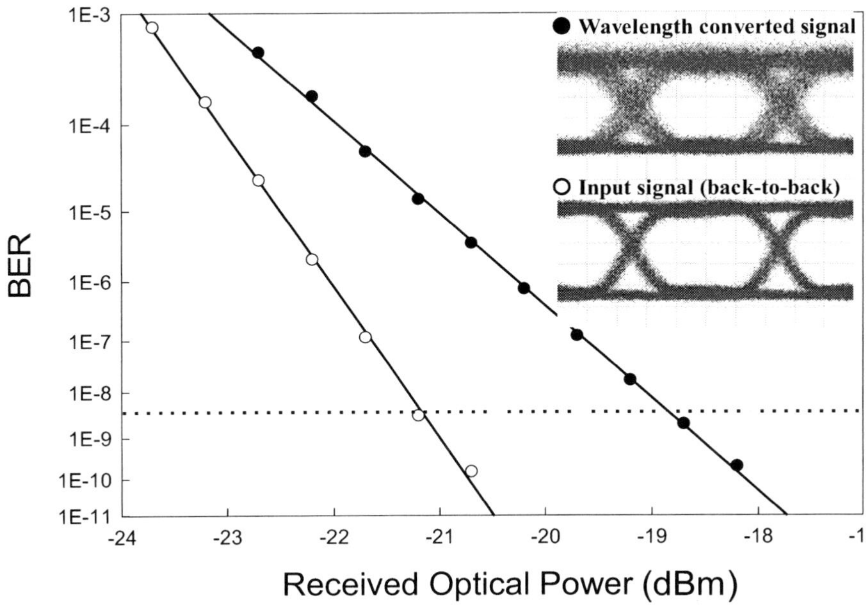

Figure 19. Measured BER versus received optical power for wavelength conversion of a 10 Gbit/s NRZ signal at a wavelength of 1544 nm. Inset: Eye diagrams. (after Ref. [35] © 2003 IEEE)

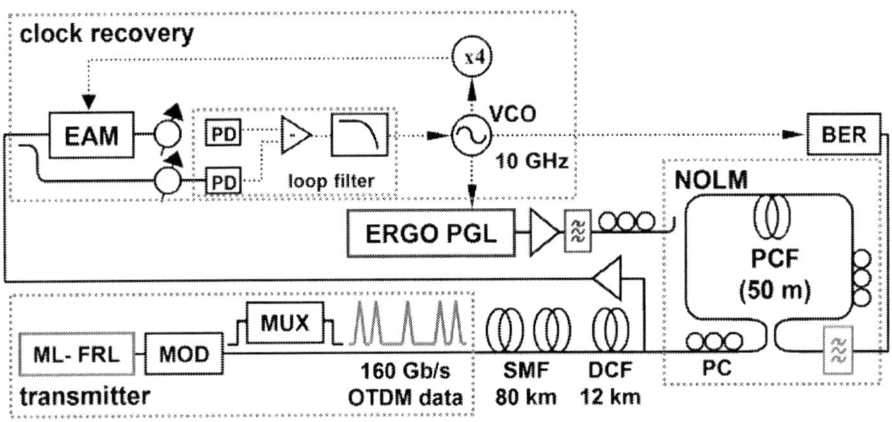

Figure 20. A schematic of a 160 Gbit/s OTDM system incorporating a HF based NOLM demultiplexer. (after Ref. [66])

4.7. OTDM Demultiplexer

Furthermore, PCF can be applied for data demultiplexing of optical time-division multiplexed (OTDM) signals, which is one of key nonlinear signal processing functions for high-speed OTDM systems. Recently, L. K. Oxenløwe et al. demonstrated a high-speed OTDM demultiplexer based on a 50m long PCF [66]. The experimental setup is shown in Figure 20. The OTDM demultiplexer was based on nonlinear optical loop mirror (NOLM) incorporating a PCF. The PCF used in this experiment had a nonlinear coefficient of 18 $W^{-1}\cdot km^{-1}$, a zero-dispersion wavelength at 1552.5 nm, and a polarization maintaining property. The pulse broadening through the PCF was less than 1%. Error-free operation at 160 Gbit/s data rate was achieved with the NOLM based demultiplexer as shown in Figure 21.

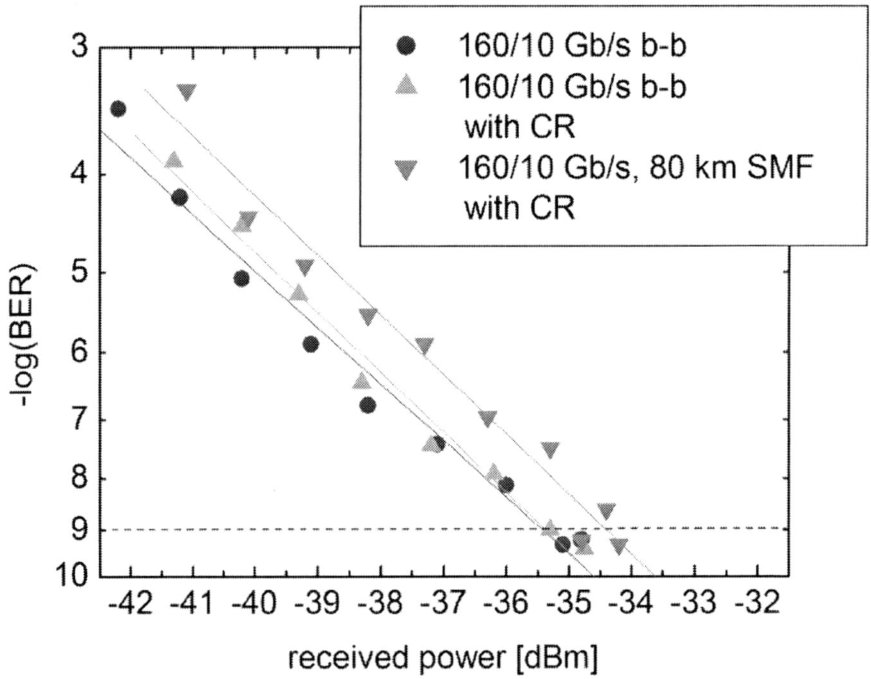

Figure 21. Measured BER and eye diagrams. (after Ref. [66])

4.8. Soft Glass-Based Highly Nonlinear PCF

Further significant increases in PCF nonlinearity can be achieved by fabricating PCFs with other glasses rather than silica, such as the Chalcogenides [49], which have around two orders of magnitude higher nonlinear optical coefficient than silica. Indeed, University of Southampton produced the first results in this direction [36] and demonstrated a PCF in SF57 lead glass (a commercial lead silica glass) with $\gamma = 640\ W^{-1}\cdot km^{-1}$, about 500 times more nonlinear than conventional SMF28 fiber [50]. This particular PCF was produced from a preform manufactured using extrusion technique [36]. The SEM image of the PCF is shown in Figure 22(a). Recently a variety of other type glass based PCFs have been demonstrated: for example, bismuth oxide [37, 67] and other lead silica glasses [68, 69]. It should be noticed that the strong waveguide dispersion of PCF allow for flexibly controlling the overall dispersion property of soft glass based PCFs at telecommunication bands in spite of the large normal dispersion associated with such materials. Raman soliton formation with a duration of 193 *fs* in 37-cm length of the SF57 lead glass-based PCF was successfully demonstrated as shown in Figure 21. (b) [50].

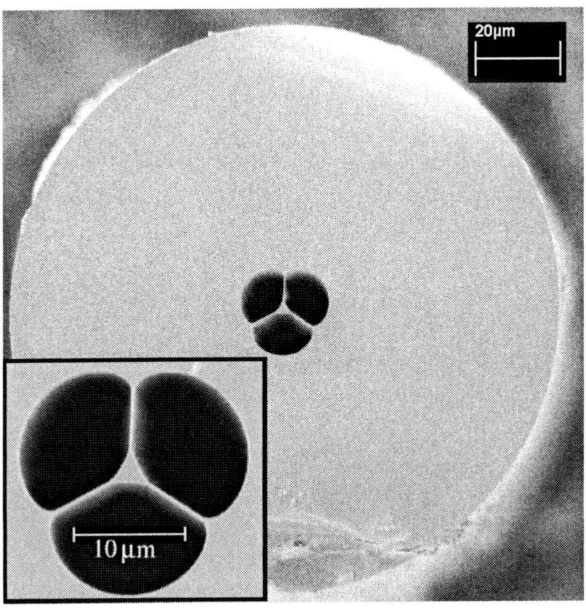

(a)

Figure 22. Continued on next page

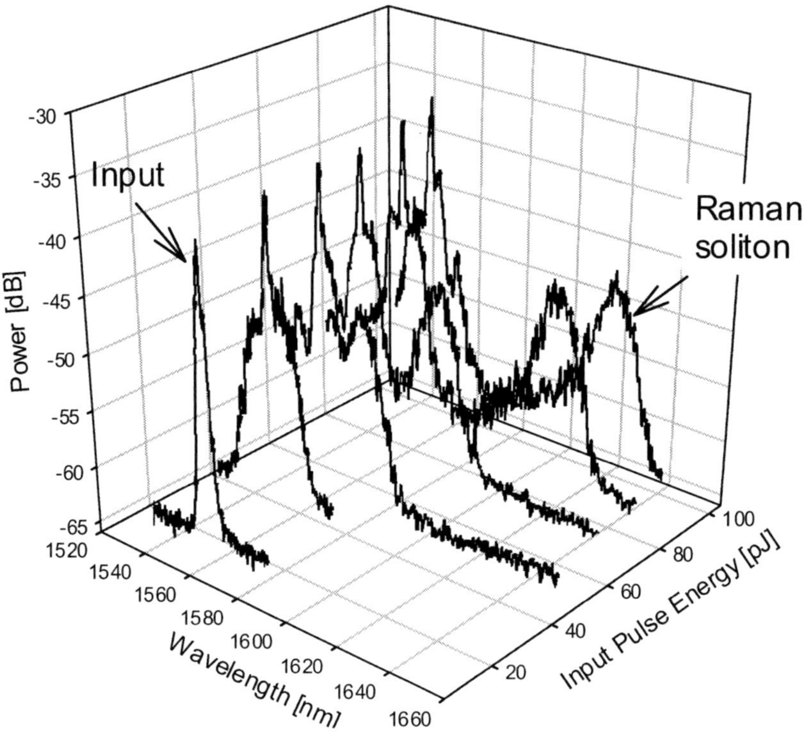

(b)

Figure 22. (a) A SEM image of a PCF in SF57 lead glass with $\gamma = 640\ W^{-1}km^{-1}$. (b) Measured pulse evolution in 37-cm length of the SF57 lead glass-based PCF vs input pulse energy. (after Ref. [50]).

Chapter 5

BISMUTH OXIDE NONLINEAR OPTICAL FIBER TECHNOLOGY FOR NONLINEAR SIGNAL PROCESSING DEVICES

5.1. BISMUTH OXIDE NONLINEAR OPTICAL FIBER (BI-NLF)

The optical fiber fabrication technology based on Bismuth oxide (Bi_2O_3) material has attracted huge attention in recent years and has resulted in the production of a variety of high quality of Bismuth oxide-based optical fibers: for example, erbium-doped Bismuth oxide-fiber (Bi-EDF) [70], highly nonlinear Bismuth oxide-fiber (Bi-NLF) [15, 71], and Bismuth oxide-based holey fiber [37, 67]. Owing to the unique material properties of Bi_2O_3 host glass a range of new possibilities, which have been considered to be beyond achievable by use of silica glass-based optical fibers can be readily offered by the Bismuth oxide-based optical fibers: for example, ultra-high erbium doping, and ultra-high Kerr nonlinearity.

Among various Bismuth oxide-based optical fibers the Bi-NLF is especially attractive since such a high nonlinearity of ~1360 $W^{-1} \cdot km^{-1}$ can be readily achieved by use of a conventional step-index fiber structure owing to an extremely-high nonlinear coefficient n_2 of the Bi_2O_3 glass [15, 70]. This means that only a meter or less in length would be long enough to generate a nonlinear optical phase shift sufficient for obtaining various nonlinear signal-processing functions. The first fabricated version of the fiber exhibited a nonlinearity parameter γ = ~ 64 $W^{-1} \cdot km^{-1}$ with an ordinary effective area A_{eff} of 20 μm^2 [70]. The Asahi Glass Company inc. then announced the fabrication of a highly advanced version of the fiber in 2003 [15] and this newly-developed Bi-NLF possessed an ultra-high nonlinearity of γ = ~1360 $W^{-1} \cdot km^{-1}$ at an effective area A_{eff} of ~3 μm^2, which is ~100 times higher than that of commercially available silica-based step index type nonlinear optical fibers. Although a step index fiber structure was employed, such an ultra-high nonlinearity could readily be obtained owing to a well defined, high quality fabrication process as well as the extremely high nonlinear coefficient n_2 of the Bi_2O_3 glass.

Figure 23(a) shows a SEM image of the Bi-NLF demonstrated in Ref. [15]. The refractive indices (n) of the core glass and the cladding glass were 2.22 and 2.13, respectively. The core and cladding glasses possess the same softening temperature and thermal expansion coefficient. The measured numerical aperture (NA) and the cut-off wavelength of the fabricated fiber were 0.64 and 1380 nm, each. The mode field diameter of this fiber was measured to be 1.98 μm and the fiber propagation loss at 1550 nm was 0.8 dB/m. Measured dispersion was -260 ps/nm-km at 1550 nm. If only one meter of the fiber is used for device implementation, the overall GVD is just -0.26 ps/nm.

Fusion splicing of newly developed optical fiber to conventional standard SMF is another important procedure from a perspective of practical fiber based device implementation. Special care must be taken of controlling arc strength and time due to the material discrepancy between Bi_2O_3 and Silica, and a high numerical aperture (NA) silica fiber is employed between them to achieve good intermediate mode mismatching. The splicing losses were measured to ~3 dB, which is acceptable for the experimental demonstration of various nonlinear signal processing devices, although more efforts are required for its further reduction.

Like other soft glass-based optical fibers the Bi-NLF also a high normal GVD at telecommunication bands due to its strong normally dispersive material dispersion. In designing an optical fiber for all-optical nonlinear signal processing, a normal GVD is essentially required at the signal wavelength since the use of anomalous dispersion fiber inevitably results in signal coherence degradation induced by modulation instability [61, 72]. Due to very high normal dispersion characteristics of the Bi_2O_3 materials at a wavelength band of ~1550 nm the GVD requirement was automatically satisfied without using any special waveguide design.

Even though the Bi-NLF can be extremely useful for the implementation of various nonlinear signal processing devices due to its huge Kerr nonlinearity and the associated short length requirement, the inherent, normal dispersion characteristic of the fiber prevent it from being employed for other types of important nonlinear signal processing devices such as parametric devices and soliton pulse generation. In order to tackle the limitation a PCF structure can be applied to the Bi_2O_3 glass as shown in Figure 23(b) of a SEM image of a fabricated 6-air hole PCF [67]. Figure 24 shows group velocity dispersion of the Bi-PCF at 1550 nm as a function of core diameter (incircle diameter: ID). Evidently, sucessful compensation of the ~-200 ps/nm/km material dispersion was readily achieved with the 6 air hole PCF structure. The overall dispersion at a wavelength of 1550 nm was measured to be between -100 and 80 ps/nm/km depending on the core diameter.

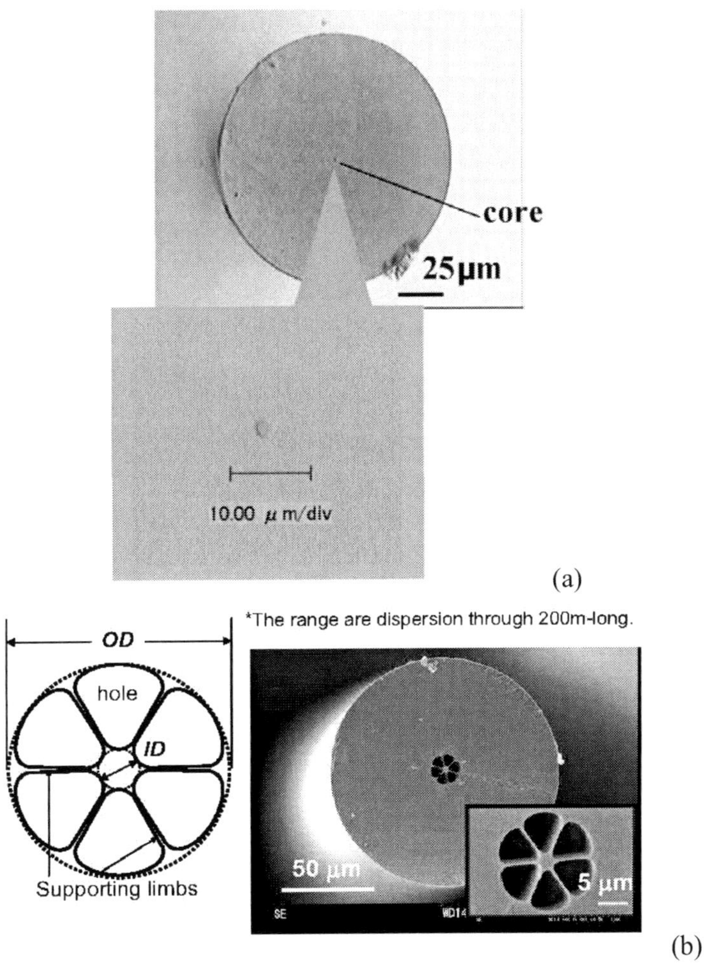

Figure 23. SEM images of (a) Bismuth oxide-based nonlinear optical fiber (Bi-NLF) (after Ref. [15]) and (b) Bismuth oxide-based PCF (Bi-PCF). (after Ref. [67])

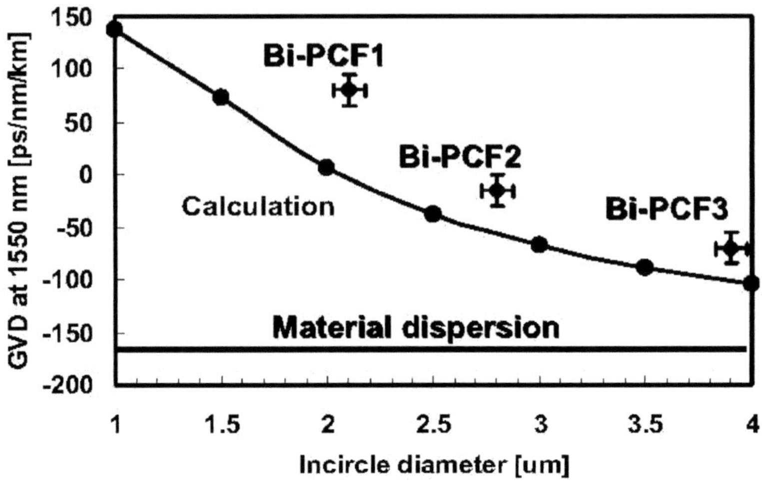

Figure 24. Measured and calculated dispersion values of the Bi-PCF shown in Figure 23(b) as a function of core size (incircle diameter). (after Ref. [67])

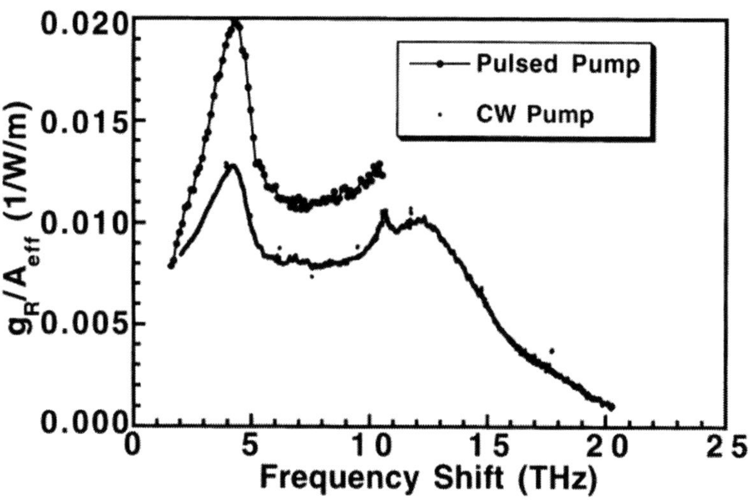

Figure 25. Measured Raman gain spectrum. (after Ref. [73])

The Raman gain coefficient measurement of a Bi-NLF was carried out by C. Cantini *et al.* using a 10-m long BI-NLF with a propagation loss of 0.5 dB/m at 1550 nm and a effective area A_{eff}= ~28 μm^2 [73]. The Bi-NLF was found to have an order of magnitude higher than that of silica-based fibers. Figure 25 shows measured Raman gain spectrum. The peak Stokes shift frequency was ~ 4 THz, which is much smaller than that of silica based fibers and the measured peak Raman gain coefficient $g_R$was ~4.2 x 10^{-13} m/W. It is noticeable that the Bi-NLF has a much flatter gain spectrum shape than that of silica-based fibers.

Furthermore, it has a unique advantage of a high SBS threshold compared to conventional silica-based nonlinear optical and proved to be superior to silica-based highly nonlinear fiber in terms of relative Kerr nonlinearity figure-of-merit considering SBS as mentioned in Chapter 3. The measured Brillouin gain coefficient g_B for a Bi-NLF with a propagation loss of 0.8 dB/m at 1550 nm and a effective area A_{eff}= ~3 μm^2 was 6.43 x 10^{-11} m/W [39]. Further details on the Bi-NLF Brillouin gain coefficient are fully described in Chapter 3.

5.2. Four-Wave Mixing-based Wavelength Converter

A practical benefit achievable from the Bi-NLF that has both an ultra-high Kerr nonlinearity and a high SBS threshold can be illustrated through implementation of a FWM wavelength converter of a NRZ signal, since this sort of optical fiber-based wavelength conversion cannot be properly performed in silica-based nonlinear fibers without SBS suppression for CW pump beams [74]. The schematic of a FWM-based wavelength converter is shown in Figure 26 [75]. A 40-cm long Bi-NLF was employed as a nonlinear medium. Note that the whole device length is almost comparable to that of the one based on fiber pigtailed semiconductor devices. Wavelength conversion of a 40-Gbit/s non-return-to-zero (NRZ) signal was performed with no additional SBS suppression scheme against the pure continuous-wave pump beam since the Bi-NLF has an extremely high SBS threshold owing to both its short length and relatively low Brillouin gain coefficient. The actual pump power within the Bi-NLF was only ~ 370 mW considering losses of the 50:50 coupler and the input splicing point, which was far below the estimated SBS threshold of 2.43 W. Figure 27(a) shows the wavelength-conversion efficiency measured as a function of the signal wavelength detuning relative to the fixed pump wavelength of 1550 nm, whereas the broken curve represents a theoretically calculated one. A maximum conversion efficiency of -15.5 dB was achieved over a 3-dB bandwidth of ~10 nm. Error-free wavelength conversion performance with a ~2 dB power penalty relative to that of the back-to-back input signal was readily achieved as shown in

Figure 27(b). No modulation instability-induced intensity noise was observed since the fiber had normal dispersion.

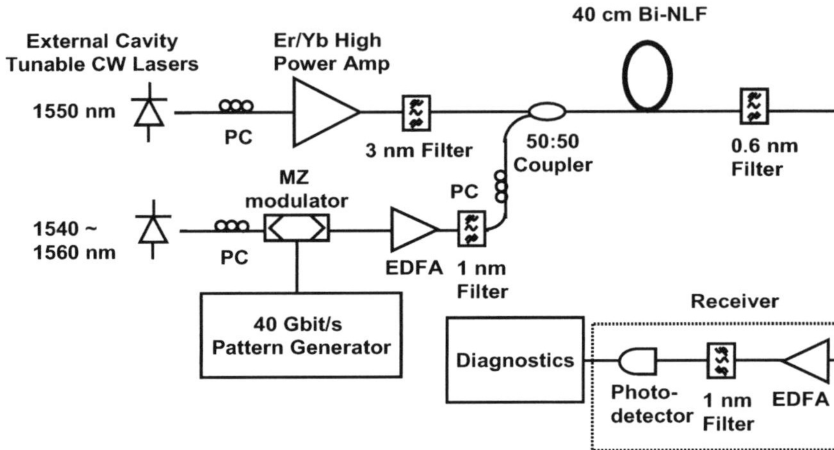

Figure 26. A schematic of FWM-based NRZ signal wavelength converter using a Bi-NLF. (after Ref. [75] © 2005 IEEE)

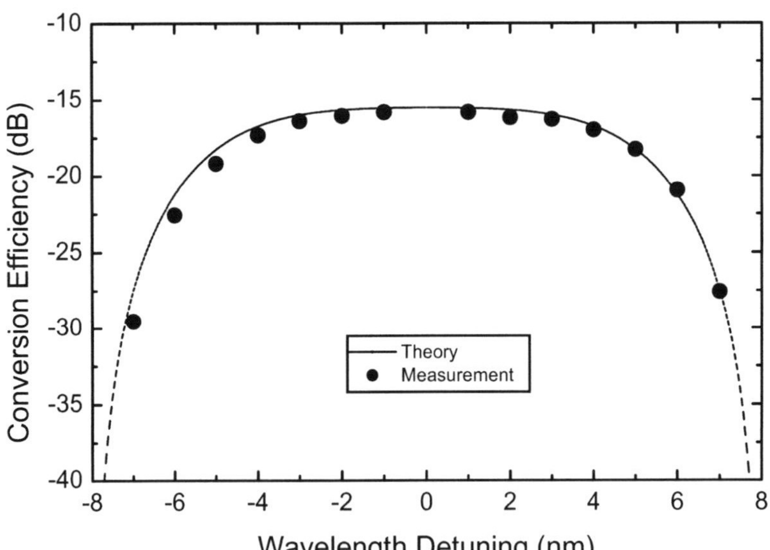

(a)

Figure 27. Continued on next page

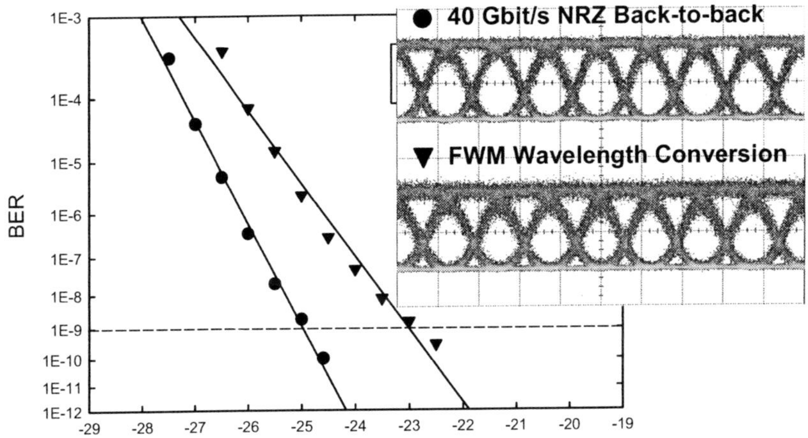

(b)

Figure 27. (a) FWM Conversion Efficiency. (b) BER and eye diagrams. (after Ref. [75], © 2005 IEEE)

5.3. High Speed OTDM Demultiplexer

Bi-NLF can also offer reduced length/power requirements for the implementation of a host of ultra-high speed OTDM nonlinear signal processing devices such as demultiplexer, add/drop multiplexer, wavelength converter, and logic gate. In order to assess the applicability of Bi-NLF to the sort of nonlinear signal processing devices a range of experimental demonstrations have been carried out. Figure 28 shows a schematic of an optical time division-multiplexed (OTDM) signal demultiplexer where the input signal was a 160 Gbit/s OTDM signal composed of 16 10-Gbit/s tributaries [76]. The demultiplexer is based on a modified Kerr shutter. In addition to nonlinear birefringence generated by the control pulses, which is employed in a conventional Kerr shutter [20], the wavelength blue shift of the data pulses induced by XPM from the control pulse trailing edge is also utilized [77]. Error-free demultiplexing of all the 16 tributaries from the 160-Gbit/s OTDM signal was achieved as shown in Figure 29.

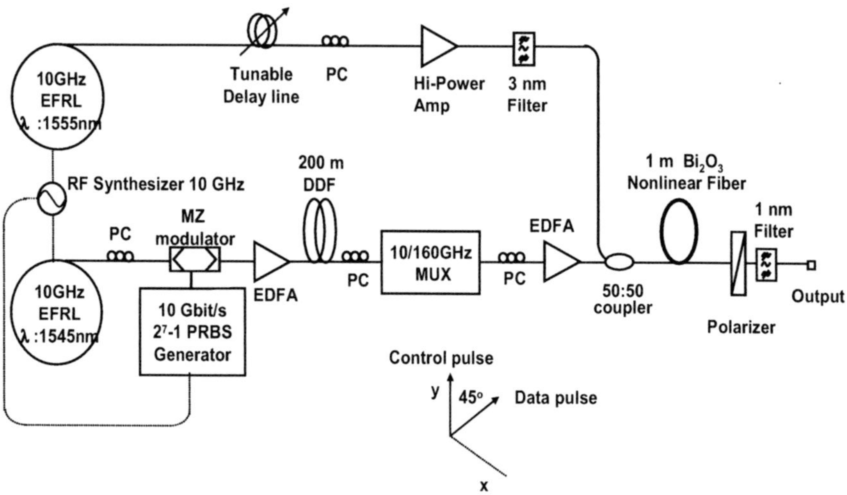

Figure 28. a schematic of a optical time division-multiplexed (OTDM) signal demultiplexer where the input signal was a 160 Gbit/s OTM signal composed of 16 10-Gbit/s tributaries. (after Ref. [76]).

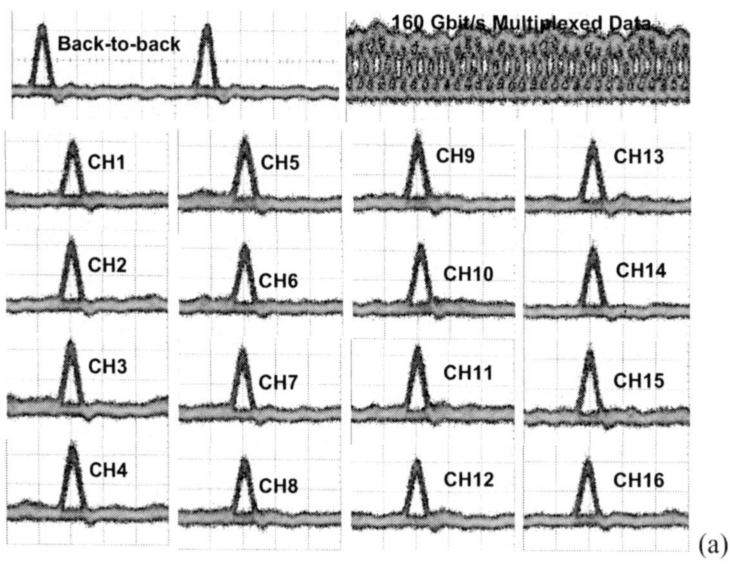

Figure 29. Continued on next page

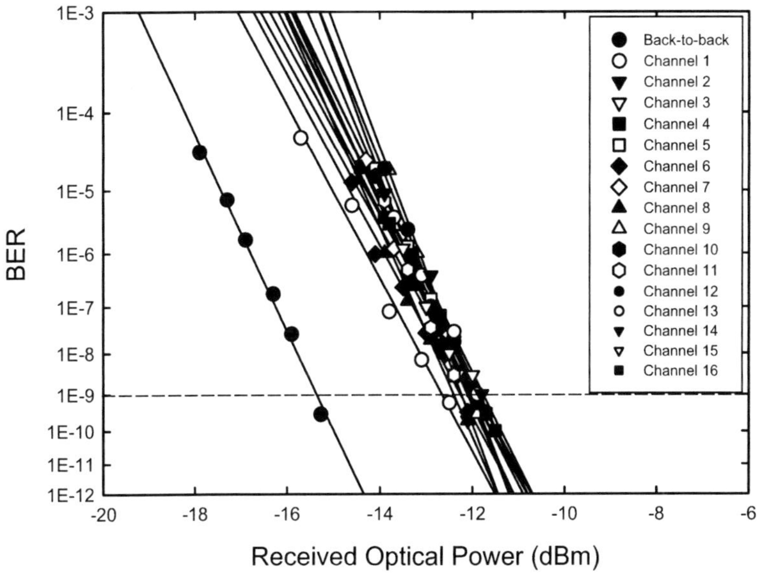

(b)

Figure 29. 160 Gbit/s OTDM signal demultipelxing. (a) Measured eye diagrams. (b) Measured BERs. (after Ref. [76])

5.4. High Speed OTDM Clock Recovery

All-optical clock recovery is another essential signal processing function in high-speed OTDM systems. A conventional way for achieving all-optical clock extraction is the use of optoelectronic phase-locked loop (OPLL) and one key issue in the OPLL-based clock recovery configuration is the optical loop length-induced timing jitter [78]. According to the previous detailed study on optical loop length-induced timing jitter [79], the long loop length caused by optical fiber amplifiers and fiber-based nonlinear phase comparators would result in a significant timing jitter, which is not tolerable for the overall system error-free operation. The use of fiber-based optical devices has thus been considered to be inappropriate for the implementation of a high-speed OTDM clock recovery.

The Bi_2O_3 optical fiber technology should be capable of solving the long-length induced limitation in implementing a high-speed OTDM receiver including both optical clock recovery and demultiplexing functions. Figure 30 shows a schematic of a high speed OTDM signal clock recovery based on Bi_2O_3 optical fiber technology [80]. The clock recovery subsystem was constructed with all fiberized components and the key components in this configuration were a Bi-NLF-based nonlinear phase comparator and a Bi-EDF-based amplifier. For the phase comparator, the cross-phase modulation (XPM)-induced polarization rotation principle was employed in a 1-m long Bi-NLF [20]. The measured switching window for this phase comparator was ~3.4 ps, which was narrow enough to resolve a 160-Gbit/s multiplexed data. The error signal was fed onto a photodetector and the generated electrical error signal was coupled into a voltage-controlled oscillator (VCO) after the offset adjustment and the lowpass filtering. The recovered 10-GHz electrical clock was then used to drive an optical clock source that is another mode-locked erbium glass laser.

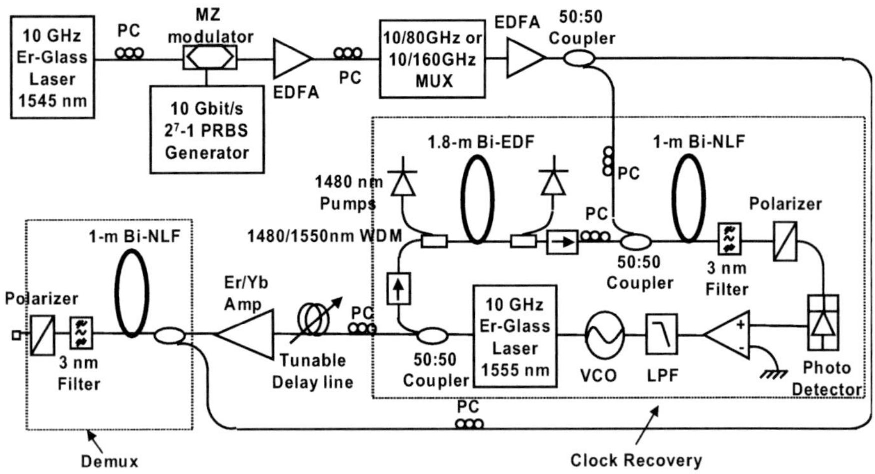

Figure 30. A schematic of a high speed OTDM signal clock recovery based on Bi_2O_3 optical fiber technology. (after Ref. [80] © 2005 IEEE)

A compact size Bi-EDFA was used to amplify the optical clock output up to a power level suitable for use of the nonlinear phase comparator. The Bi-EDFA was composed of a 1.8-m long Bi-EDF with an erbium concentration of 3500 ppm. A clean frequency peak of the recovered electrical clock from the 160-Gbit/s multiplexed PRBS data stream was obtained as shown in Figure 31. Its root mean square (RMS) timing jitter was estimated to be ~520 fs. Figure 32 shows measured eye diagrams and the corresponding BERs. Error-free demultiplexing with the recovered clock was readily achieved with a ~2-dB power penalty relative to demultiplexing with the transmitter clock at 80 Gbit/s.

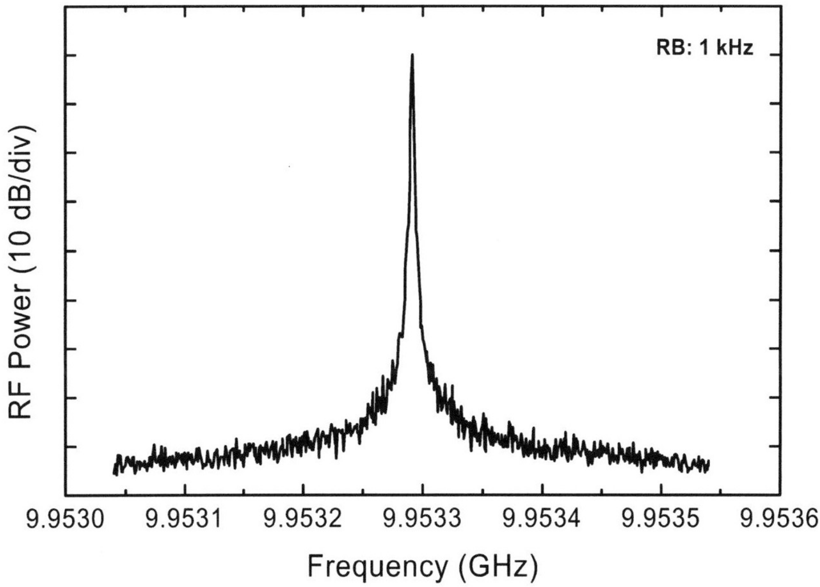

Figure 31. Measured RF spectrum of the recovered clock. (after Ref. [80] © 2005 IEEE)

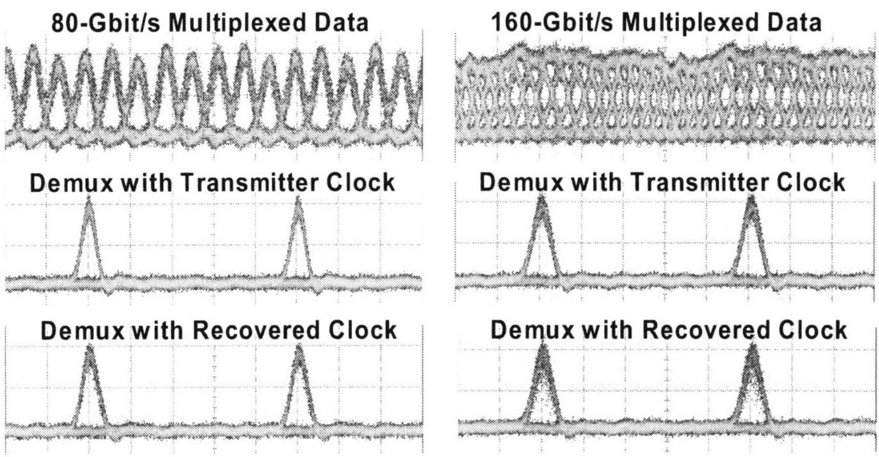

(a)

Figure 32. Continued on next page

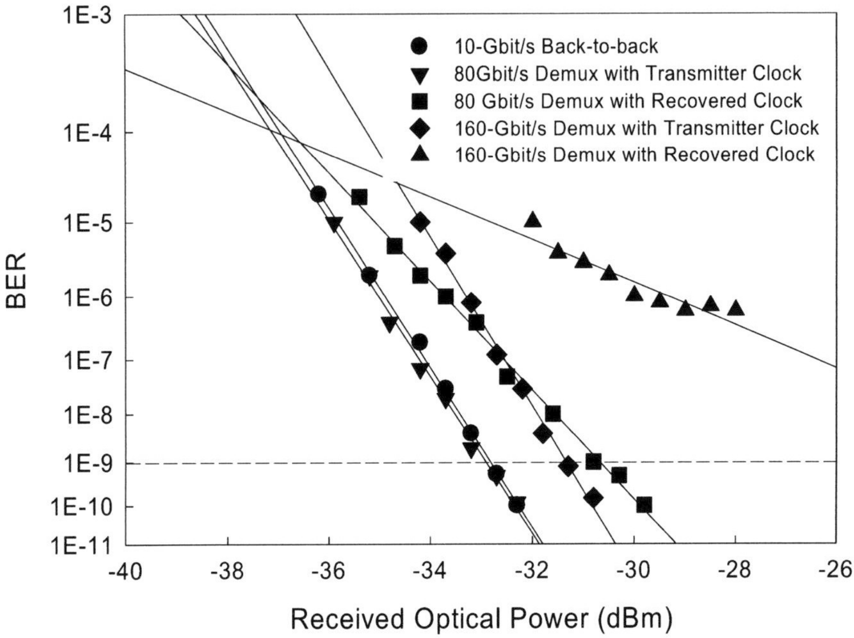

(b)

Figure 32. Measured (a) eye diagrams and (b) BERs versus received optical power for the demultiplexed signals. (after Ref. [80] © 2005 IEEE)

5.5. 160 Gbit/s Wavelength Converter

Figure 33(a) shows an experimental schematic of a 160-Gbit/s OTDM signal wavelength converter employing a 1-m long Bi-NLF [81, 82]. The wavelength converter is based on the conventional Kerr shutter using nonlinear birefringence [20]. XPM between the control and the CW beams results in nonlinear birefringence for the CW laser beam where these beams overlap temporally within the fiber. This nonlinearly polarization-rotated signal can then be converted to a wavelength-converted signal by passing it through a polarizer, which serves to eliminate the residual CW signal with no polarization rotation [83]. In order to implement a stable and high performance Kerr shutter using a conventional optical fiber, a suppression scheme for signal polarization instability induced by temperature-dependent local birefringence fluctuation owing to its long fiber length, is essentially required [84]. However, the use of a very short, ultra-high nonlinearity fiber could significantly reduce the polarization instability problem [83].

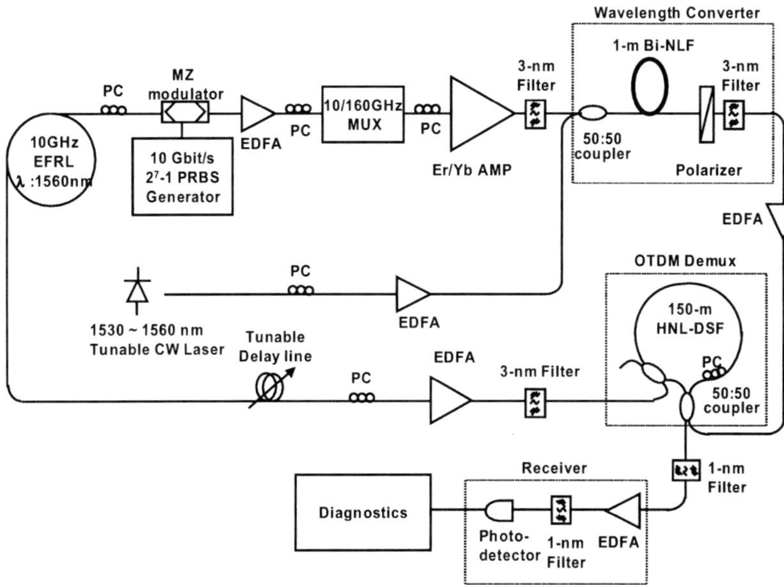

(a)

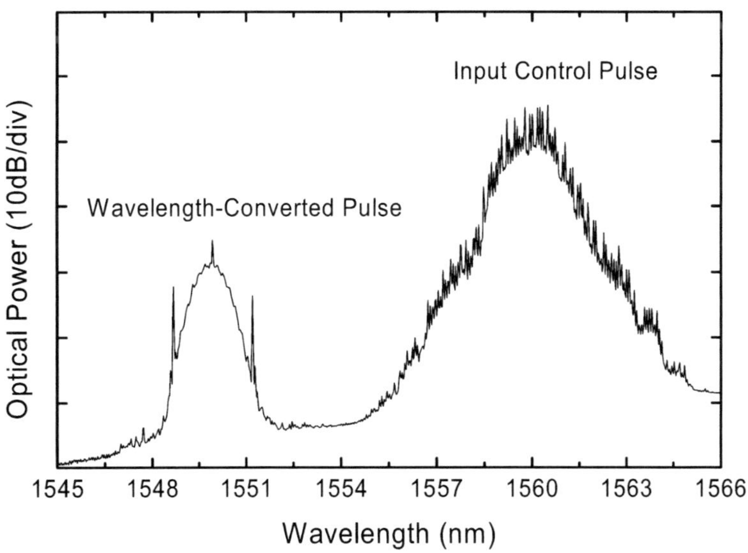

(b)

Figure 33 (a) A schematic of a 160-Gbit/s OTDM signal wavelength converter based on a 1-m long Bi-NLF. (b) Measured optical spectrum after the polarizer. (after Ref. [81])

The output optical spectrum measured after the polarizer is shown in Figure 33(b). The CW probe beam was readily converted into a 160-Gbit/s signal due to nonlinear birefringence induced by the control beam. High spectral quality of the wavelength-converted signal is clearly evident from the graph. The conversion performance was quantified with BER after demultiplexing the 160-Gbit/s wavelength-converted pulses to 10-Gbit/s tributaries, and error-free operation was readily achieved with a power penalty of ~6.5 dB relative to the back-to-back.

5.6. 160 Gbit/s OTDM Add/Drop Multiplexer

An essential signal processing device in high-speed OTDM systems and the related all-optical networks is the channel add/drop multiplexer (ADM) operating in the time domain [85]. The ADM performs two key functions: One is selectively dropping a base-rate tributary from a multiplexed high bit-rate data stream in the time domain, and the other is adding a new channel into the cleared dropped-bit time slot [86] at the network nodes. Two complementary switching windows with a high-extinction ratio are thus required for simultaneously realizing good add and drop performance. A variety of OTDM ADMs have been proposed and experimentally demonstrated to date and those are based on optical switches incorporating either semiconductors devices [85, 86] or nonlinear optical fibers [87, 88, 89]. The applicability of our Bi-NLF technology for the implementation of ultra-high speed, nonlinear signal-processing devices was investigated [90]. Figure 34 shows a schematic of a 160-Gbit/s OTDM channel add/drop multiplexer using only 1-m length of Bi-NLF.

The OTDM add/drop multiplexer is based on XPM-induced nonlinear polarization rotation [20, 91]. XPM between the control and the multiplexed data beams results in nonlinear birefringence for the data beam where these beams overlap temporally within the fiber. Only the nonlinearly polarization-rotated channel (drop-channel) by the synchronized control pulse passes through one output port of the polarization beam splitter and is dropped, whereas the rest of the channels with a cleared drop-channel time slot (through-channels) are extracted from the other port of the beam splitter. A new channel (add-channel) is then added to the empty time slot of the reflected through-channels using a fiber coupler.

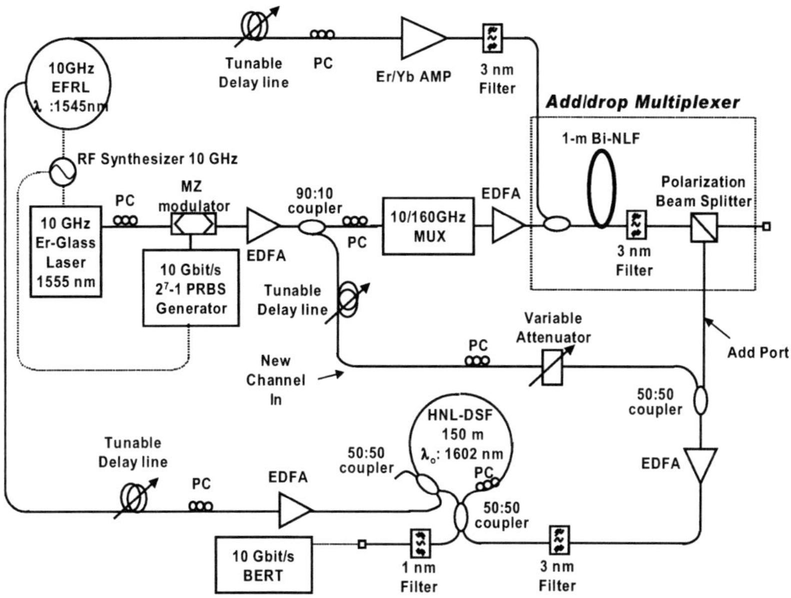

Figure 34. A schematic of a 160-Gbit/s OTDM channel add/drop multiplexer using only 1-m length of Bi-NLF. (after Ref. [90])

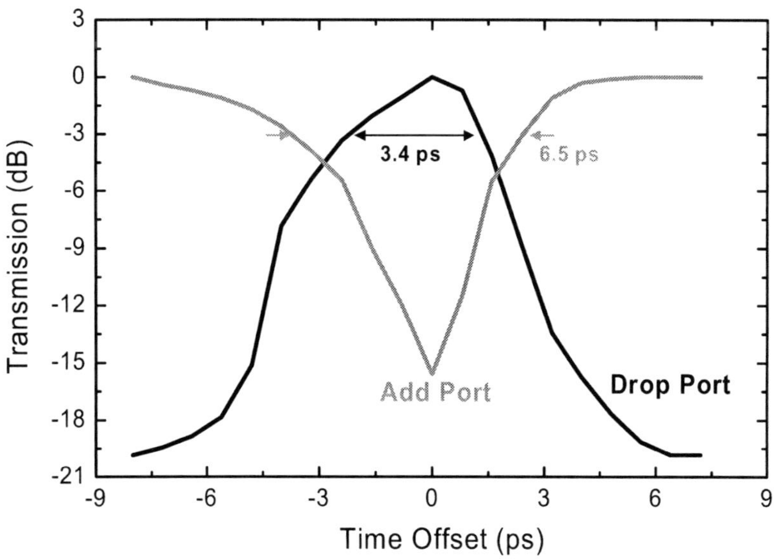

Figure 35. Measured switching windows for add and drop ports of our 1-m Bi-NLF based ADM. (after Ref. [90])

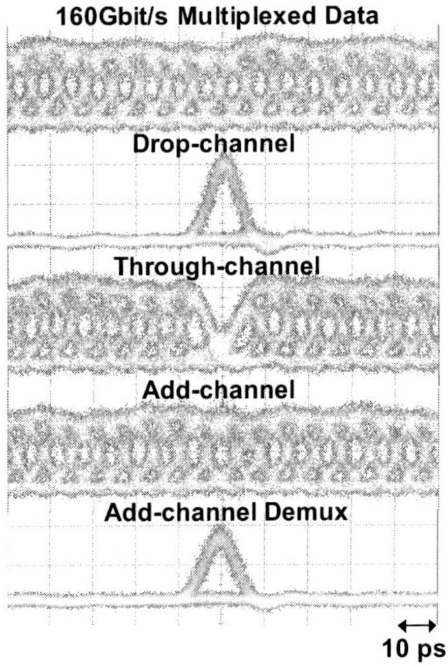

(a)

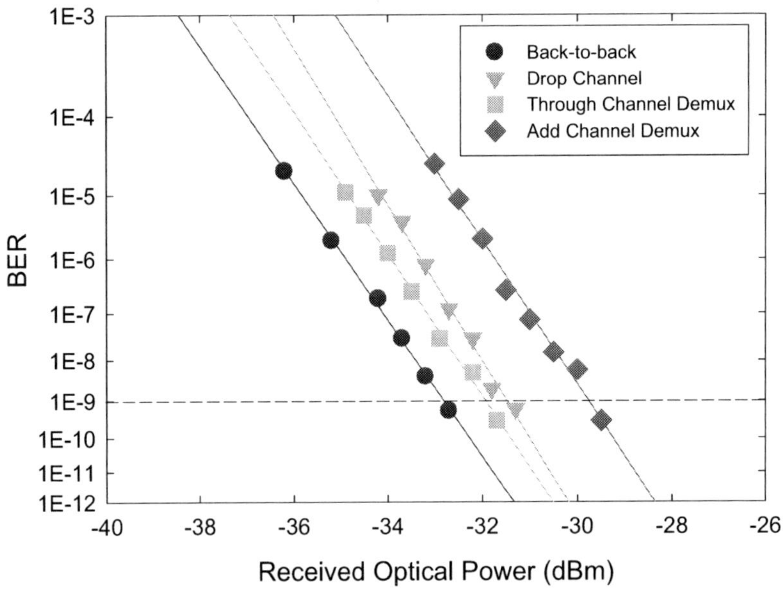

(b)

Figure 36. (a) Measured eye diagrams and (b) BERs. (after Ref. [90])

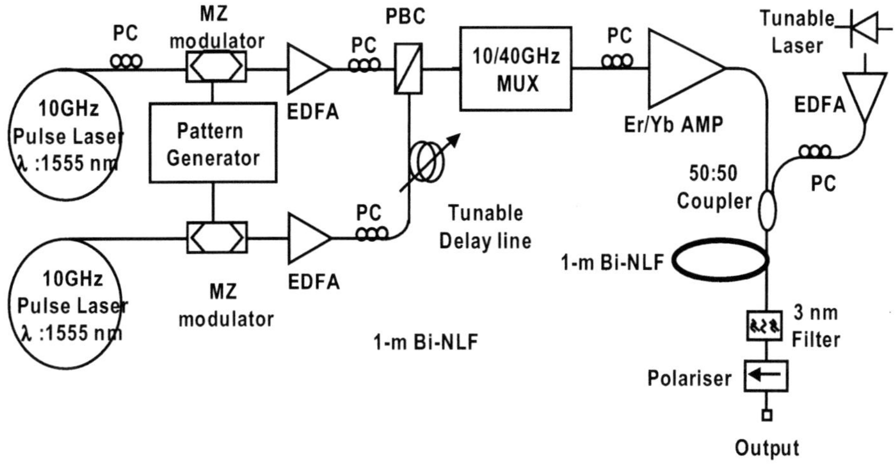

Figure 37. A schematic of XOR and AND gates using a-m long Bi-NLF. (after Ref. [95])

Figure 35 shows the measured switching windows of both the drop and the add ports by sliding the control pulse across the data pulse in the time domain and measuring the output optical power. The full width at half maximum (FWHM) of the measured windows was 3.4 ps and 6.5 ps for the drop and the add ports, respectively. High on-off extinction ratios of ~20 dB and ~16 dB were achieved for the drop and the add ports, each.

Figure 36(a) shows measured eye diagrams for the input OTDM channels, the drop-channel, the through-channels, the newly reconstructed OTDM channels after the ADM, and the demultiplexed add-channel, which were measured using a fast p-i-n diode and sampling oscilloscope of a combined ~70 GHz bandwidth. High quality of simultaneous add-drop operation is clearly evident from the eye diagrams. The ADM showed error-free operation at a multiplexed data rate of 160 Gbit/s as shown in Figure 36(b).

5.7. 40-Gbit/s Logic Gates of AND/XOR

All-optical logic gates are key devices for building future high capacity, all-optical networks. The logic gates of XOR, AND, NAND, and INVERSION perform a range of essential signal processing functions at the network nodes such as header processing, pattern matching, encryption, and contention resolution

[92]. All-optical logic gates can be implemented by use of nonlinear optical effects within semiconductors [93] or optical fiber [94]. Figure 36 shows a schematic of XOR and AND gates using a 1-m long Bi-NLF [95].

XOR

P1	P2	Out
1	1	0
1	0	1
0	1	1
0	0	0

AND

P1	P2	Out
1	1	1
1	0	0
0	1	0
0	0	0

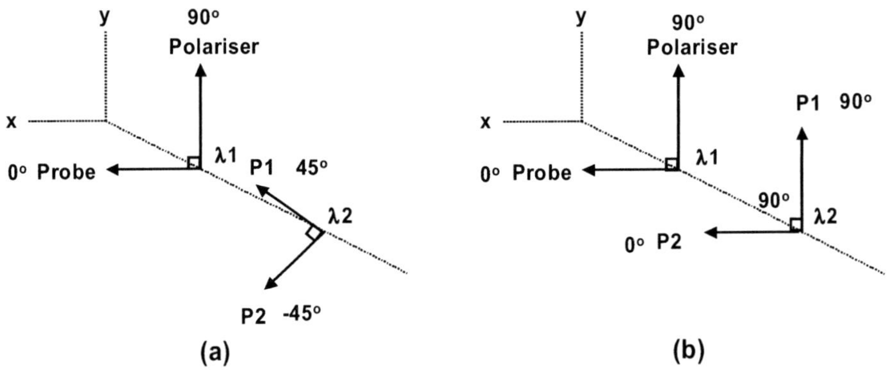

Figure 38. Operating principle of XOR and AND gates. (after Ref. [95])

The operating principle of the schematic is as follows. The XOR gate is based on XPM-based nonlinear polarization rotation, the basic concept of which was proposed by C. Yu *et al.* [94]. The SOPs of the two input signal beams (P1 and P2) are orthogonal each other at the input of a Bi-NLF and the probe beam polarization is aligned to have a 45^{0} angle with respect to both input beams as shown in Figure 38(a). A polarizer is located at the output to be orthogonal to the probe polarization. When either P1 or P2 exists, XPM-induced polarization rotation of the probe beam occurs and the corresponding spectral components pass through the polarizer. However, when both P1 and P2 coexist, nonlinear polarization rotation caused by P1 cancels out that by P2, and no output is obtained after the polarizer. One more interesting point in this particular configuration is that we can also obtain the AND function within the same device by simply changing the SOPs of the two input beams by 45^{0} as shown in Figure

38(b). In this case, the SOP of P1 is parallel to the polarizer whilst the SOP of P2 is parallel to that of the probe beam. Since the SOPs of the neighboring beams are perpendicularly polarized, orthogonal-polarized wavelength exchange occurs [56]. An output signal can thus be obtained out of the polarizer only when both P1 and P2 exist within the fiber. Note that phase mismatch due to walk-off effect is negligible in such short length of fiber.

Figure 39 shows oscilloscope traces for XOR and AND operation at 40 Gbit/s. It is clearly evident from the graphs that good performance of XOR and AND operation was readily achieved in such a compact, single fiber-based device.

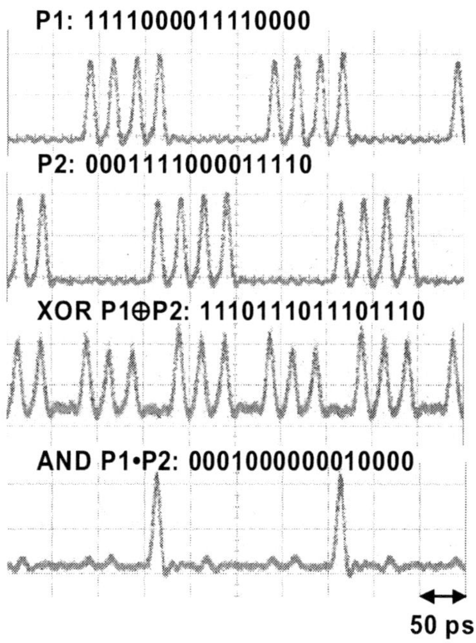

Figure 39. Measured oscilloscope traces. (after Ref. [95])

5.8. Optical Phase Conjugator

Midway Optical phase conjugation (OPC) has attracted huge technical interest and has been investigated as a promising technique for restoration of signal distortion caused by chromatic dispersion, nonlinear effect, or both in optical communication systems [96]. For achieving the signal restoration in a transmission link a phase conjugator that inverts an input signal in the spectral domain is usually located at a mid point of the transmission link. Thus, chromatic dispersion and/or nonlinear impairments experienced by the signal in the front half of the transmission link are automatically compensated after the second half transmission. This technique is also called Mid Span Spectral Inversion (MSSI).

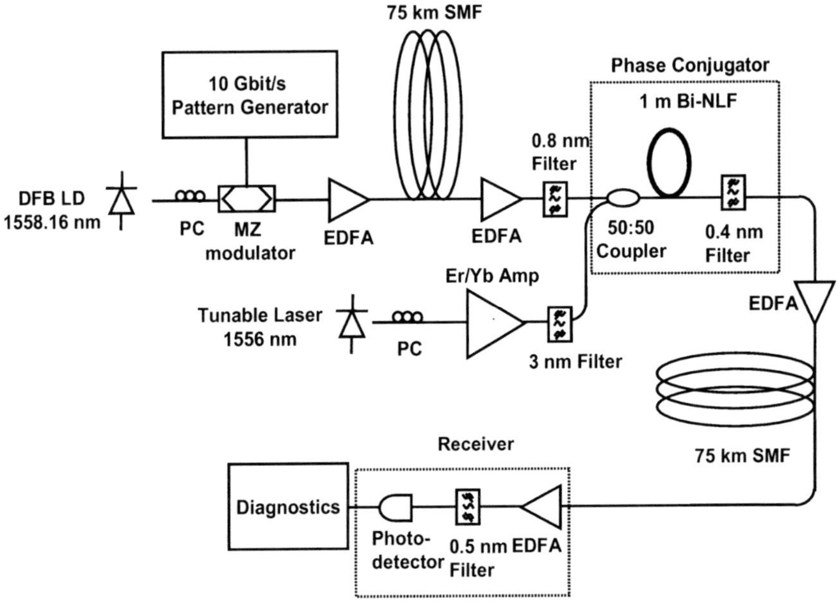

Figure 40. A schematic for 150-km SMF transmission of a 10-Gbit/s NRZ signal with a Bi-NLF based phase conjugator. (after Ref. [103])

In the OPC technique the key optical device is an all-optical phase conjugator. A phase conjugate of a signal is obtained through the use of either the third order nonlinearity $\chi^{(3)}$ associated four-wave mixing in semiconductor optical amplifiers (SOAs) [97] and optical fiber [98, 99], or the cascaded second order nonlinearity ($\chi^{(2)}$: $\chi^{(2)}$) in periodically-poled $LiNbO_3$ (PPLN) waveguides [100]. Although optical fiber has a range of advantages over SOAs and PPLNs such as no optical signal-to-noise-ratio (OSNR) limitation caused by amplified spontaneous emission that is a problem in SOAs [101] and no need for precise, high temperature control that is required for stable phase matching of PPLNs [100], it has been less favourite than SOAs and PPLNs from a perspective of practical device implementation due to SBS induced low conversion efficiency [102], the corresponding OSNR degradation of the output phase conjugate, and the long length associated stability and compactness issue. The three issues inherent to optical fiber based phase conjugator can readily be sorted out by using Bi-NLF.

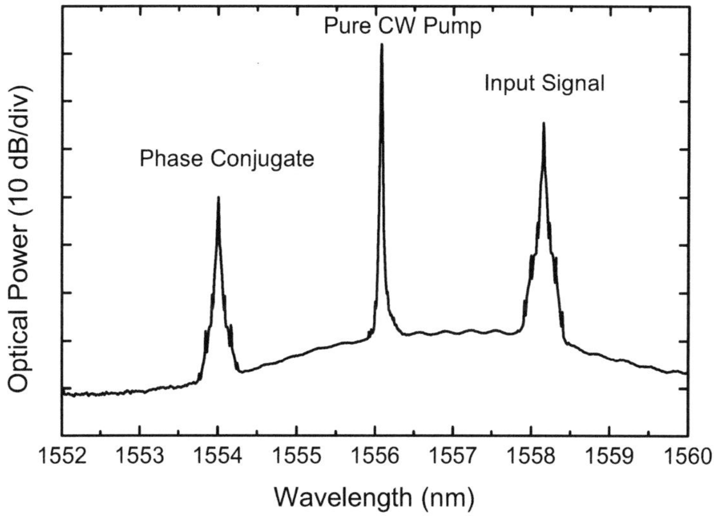

(a)

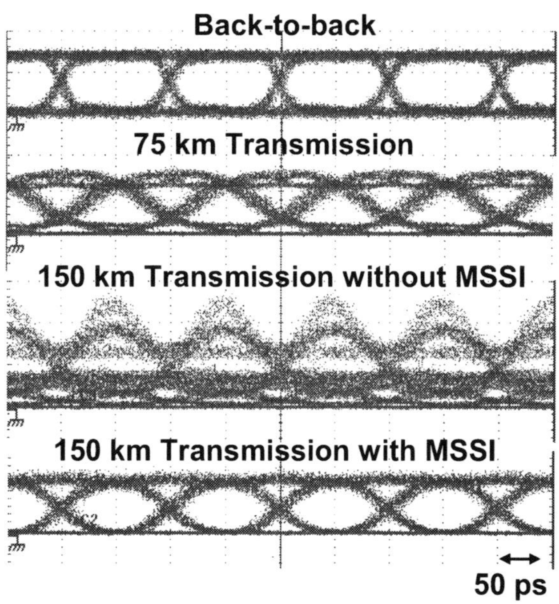

(b)

Figure 41. (a) Measured optical spectrum at the output end of the Bi-NLF. (b) Measured eye diagrams. (after Ref. [103])

Figure 40 shows a schematic for 150-km SMF transmission of a 10-Gbit/s NRZ signal with a Bi-NLF based phase conjugator [103]. A phase conjugate of the transmitted signal was generated through FWM within the Bi-NLF.

Figure 41(a) shows the measured optical spectrum at the output end of the Bi-NLF. The OSNR of the output phase conjugate was ~31 dB whilst that of the input 75-km transmitted signal was ~40 dB, indicating high quality of optical phase conjugation in the spectral domain. The conversion efficiency, which was defined as the ratio of the input signal power and its phase conjugate power was estimated to be ~ -14.5 dB. The values of the output OSNR and the conversion efficiency are not straightforward to obtain by use of a conventional silica based highly nonlinear fiber with no SBS suppression scheme.

Figure 41(b) shows measured eye diagrams. The 10-Gbit/s signal was significantly distorted even after the 75-km transmission and serious distortion of the signal was observed after transmission over 150 km without dispersion compensation as expected. In this case BER measurement was impossible to be performed. However, almost perfect signal restoration could be achieved when the Bi-NLF based phase conjugator was employed at the mid point of the 150 km transmission link. Using the phase conjugator, error-free signal detection with a power penalty of ~1dB was successfully achieved after transmission over the 150 km SMF.

5.9. 2R Regenerator

All-optical data regenerators that can restore the degraded quality of transmitted signals are of great practical interest for the improvement of both overall system performance and cost effectiveness in current ultra-long haul transmission systems. All-optical regenerators can be classified as 2R or 3R types depending on whether the retiming function is incorporated into the devices that perform the basic two functions of reshaping and reamplification. A range of all-optical regenerators have been proposed and demonstrated so far by use of semiconductor optical amplifiers (SOAs) [104], electroabsorption modulators (EAMs) [105], or nonlinear optical fiber [11, 106, 107].

One of the most attractive regeneration schemes is the Mamyshev type regenerator that is based on self-phase modulation induced spectral broadening in nonlinear optical fiber and subsequent spectral filtering [11, 106, 107]. Due to its structural simplicity, polarization insensitivity, and fast response time (a few fs), the Mamyshev type regenerator has been considered to be a highly promising approach for the replacement of the conventional regenerators that require the fast conversion between the optical and the electrical domains although it has inherently 2R regeneration characteristics. It was also proposed that 3R operation of the Mamyshev type regenerator is possible through the combined use of a synchronous modulator [108].

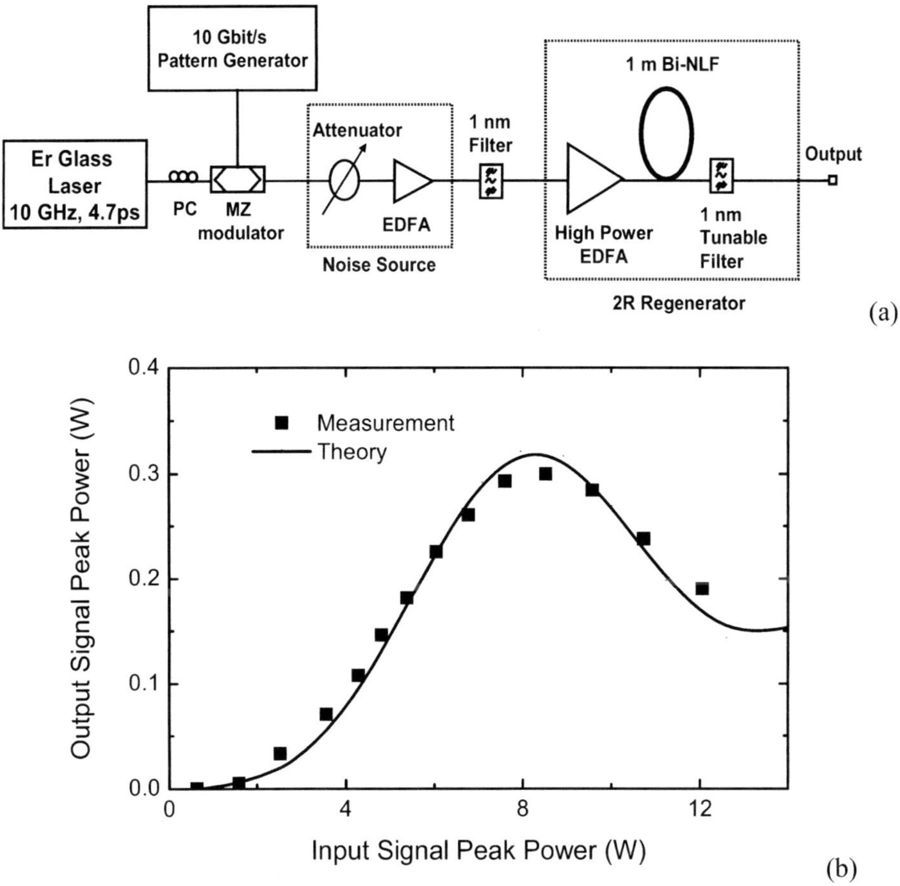

Figure 42. (a) A schematic for Mamyshev type 2R regeneration based on a 1-m long Bi-NLF. (b) Experimentally measured and calculated power transfer characteristic of the Bi-NLF based Mamyshev type 2R regenerator as a function of the input signal peak power. (after Ref. [111] © 2006 IEEE).

In the design and practical implementation of a high performance Mamyshev type regenerator the most critical component is a nonlinear optical fiber. In order to obtain an ideal, step-like power transfer function, the nonlinear fiber used within the regenerator should satisfy the tight parameter requirements such as low group velocity dispersion (GVD), normal dispersion, high $\chi^{(3)}$ nonlinearity, and high stimulated Brillouin scattering (SBS) threshold [108]. So far, km length of silica based highly nonlinear dispersion-shifted fiber (HNL-DSF) has been employed for most of the regenerator implementations [106, 107, 108, 109]. Although the silica based HNL-DSF allows for good performance data regeneration, the stability and compactness issues caused by the long fiber length still remain challenging. In order to tackle the issues, Fu *et al.* has proposed the use of a soft glass based ultra-high nonlinearity optical fiber that is As_2Se_3 Chalcogenide fiber [110]. The feasibility of the use of our fabricated Bi-NLF to implement a Mamyshev type 2R regenerator has also been investigated [111].

Figure 42(a) shows a schematic for Mamyshev type 2R regeneration based on a 1-m long Bi-NLF. In Figure 42(b), the experimentally measured power transfer characteristic of the Bi-NLF regenerator is plotted as a function of input signal peak power. A good S-shape transfer characteristic for the regenerator is clearly shown in the graph, implying that both mark and space noise on the input signal can be readily suppressed by setting the input signal peak power at an optimum operating point.

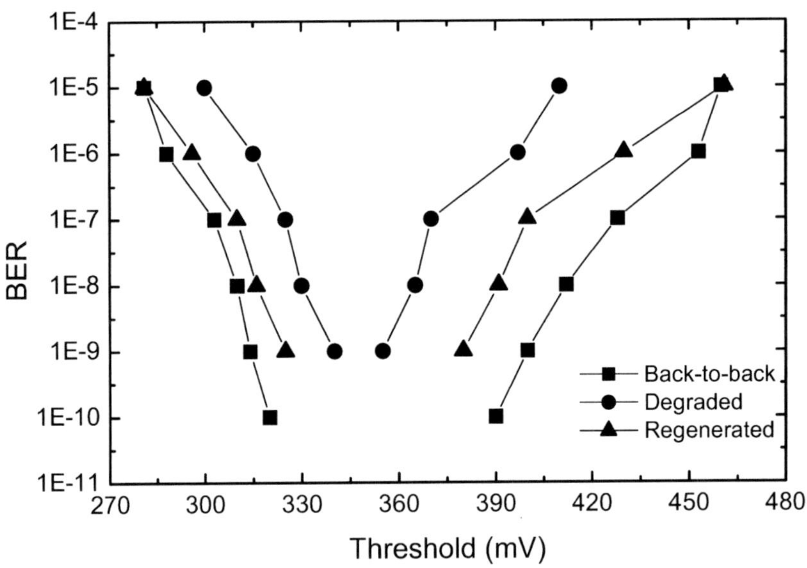

Figure 43. Measured BERs while the decision threshold level was changed at a fixed receiver optical power of -28 dBm. (after Ref. [111] © 2006 IEEE)

Figure 43 shows measured BERs for the three cases i.e. the signal back-to-back, the degraded signal, and the regenerated signal while the decision threshold level was changed at a fixed receiver optical power of -28 dBm. For this measurement, a p-i-n receiver preamplified by an EDFA with a ~3.5 dB noise figure at 1550 nm was used. The mark and space intensity noise were significantly reduced after the regenerator. The degree of the mark and space noise suppression was quantified by estimating Q-factors for the three cases [112]. The Q value of the regenerated signal was estimated to be ~17.1 dB while that of the degraded signal was ~15.8 dB. The back-to-back Q was ~18 dB. A Q-factor improvement of 1.3 dB was thus achieved after the regeneration.

5.10. XPM-Based NRZ Signal Wavelength Converter

NRZ signal WC within nonlinear optical fiber can be achieved by use of either four-wave mixing (FWM) [74] or cross phase modulation (XPM) [113]. Although XPM is less sensitive to the group-velocity dispersion (GVD)-induced walk-off effect and allows for wideband wavelength tuning, XPM-based WC has attracted less attention than FWM WC since it has a lower conversion efficiency [113]. There are two issues to be sorted out in realizing XPM-based wideband NRZ wavelength converters [113, 114]. One is the high peak power requirement for the control signal (modulated data signal) [113]. Due to a high duty cycle an extremely high average power of the control signal is required to generate a sufficient nonlinear phase shift for the conversion operation unless an ultra-high nonlinearity fiber is employed. The other is concerning how to implement a SBS suppression scheme [35, 64]. Note that the commonly used frequency dithering scheme directly results in the converted signal spectrum broadening, which is not acceptable in real systems [115]. The simplest way to tackle the two issues is to use an ultra-high Kerr nonlinearity fiber with a high SBS threshold. Bi-NLF should thus provide an excellent for the issue. Figure 44 shows a schematic of NRZ signal wavelength converter based on XPM in optical fiber [116].

The wavelength converter is based on the XPM-induced nonlinear polarization rotation principle [20, 116]. In case of an NRZ signal we can obtain pattern-inverted or non-inverted wavelength conversion depending on the relative states of polarization (SOPs) between the probe beam and the polarizer. When the SOPs of the probe beam and the polarizer are orthogonal to each other and the input signal polarization is aligned to have a 45° angle with respect to that of the polarizer as shown in Figure 45, we obtain a pattern-non-inverted output signal after the polarizer. This configuration is well known as the Kerr shutter [20] and has been commonly used in wavelength conversion or demultiplexing of RZ signals [117]. However, when the SOPs of the probe and the polarizer are parallel to each other, a pattern-inverted signal is obtained. Note that the pattern inversion can occur only if the input has a NRZ format. Compared to the previous XPM-based NRZ signal wavelength converters using optical fiber nonlinear loop mirror [113, 114], the scheme is simple and does not require any additional scheme for solving the extinction ratio degradation problem caused by the counter-propagating beam-induced undesired nonlinear phase shift.

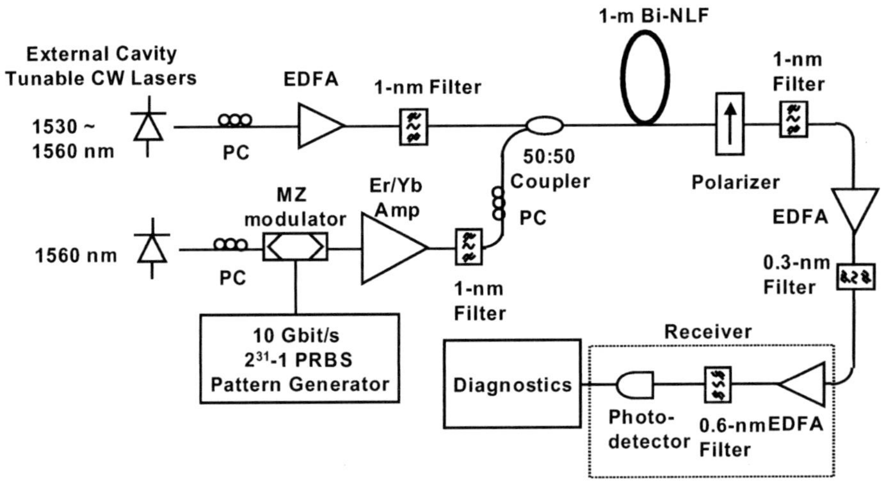

Figure 44. A schematic of NRZ signal wavelength converter based on XPM in optical fiber. (after Ref. [116] © 2006 IEEE)

Figure 46(a) shows the optical spectrum at the output end of the Bi-NLF. High quality of WC in the spectral domain was clearly obtained for both cases, i.e. non-inverted and inverted patterns from the close-up views. A conversion efficiency, which was defined as the ratio of the output wavelength converted signal power and the input signal power was estimated to be ~ -26 dB. Figure 46(b) shows measured eye diagrams. Error-free wavelength conversion over a 30-nm bandwidth was readily achieved for both cases, i.e. non-inverted and inverted patterns. Interestingly, the pattern-inverted WC was found to have better BER performance (~2 dB penalty) than the pattern-non-inverted one (~5.5 dB penalty).

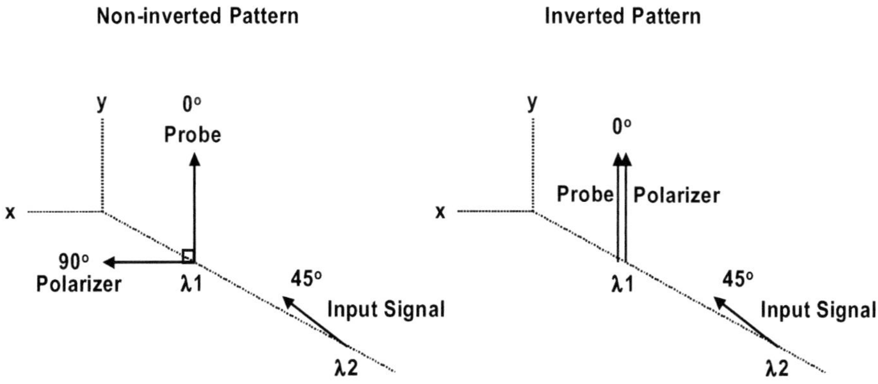

Figure 45. Operating principle of XPM-induced nonlinear polarization rotation- based wavelength conversion. (after Ref. [116] © 2006 IEEE)

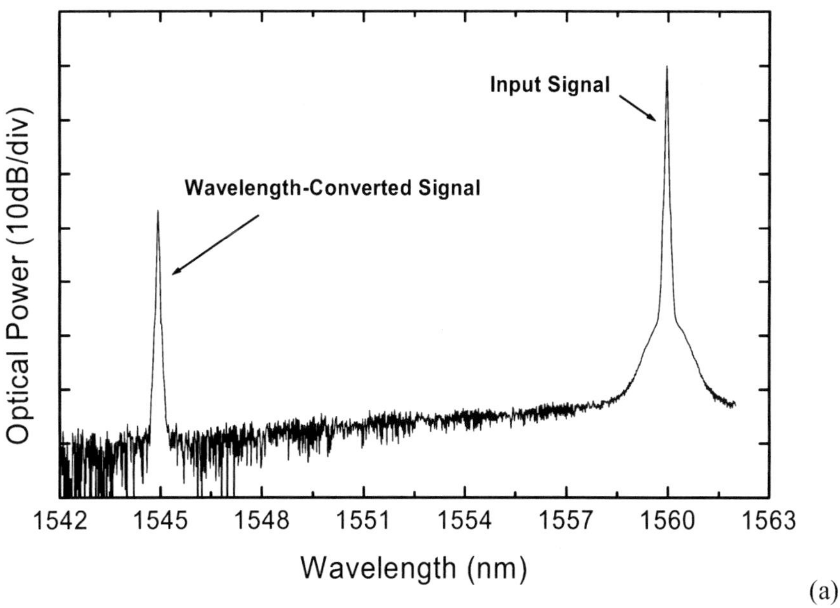

Figure 46. Continued on next page

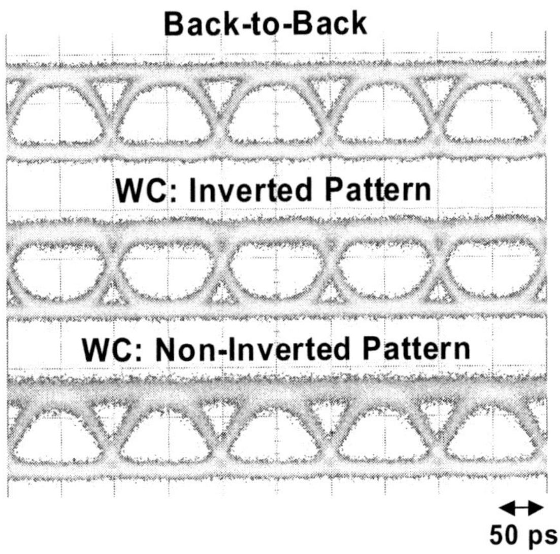

(b)

Figure 46. (a) Optical spectrum at the output end of the Bi-NLF. (b) Measured eye diagrams. (after Ref. [116] © 2006 IEEE)

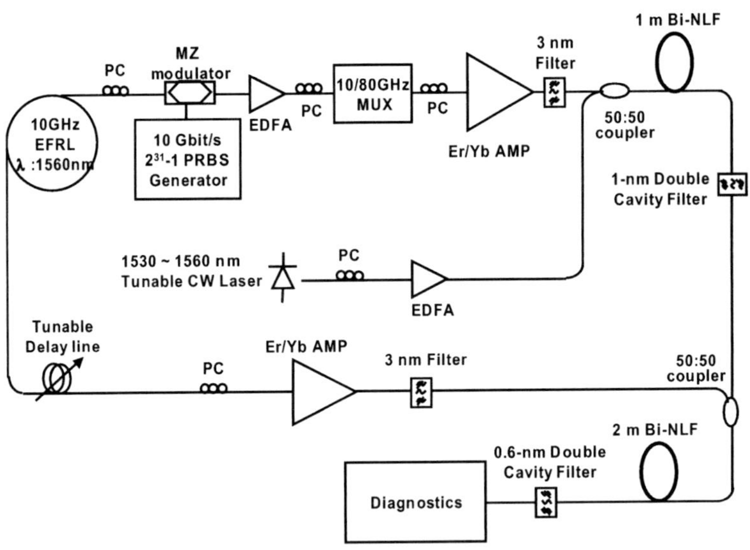

Figure 47. A experimental schematic for wavelength conversion and demultiplexing based on the XPM and subsequent offset filtering principle in short lengths of Bi-NLF. (after Ref. [118])

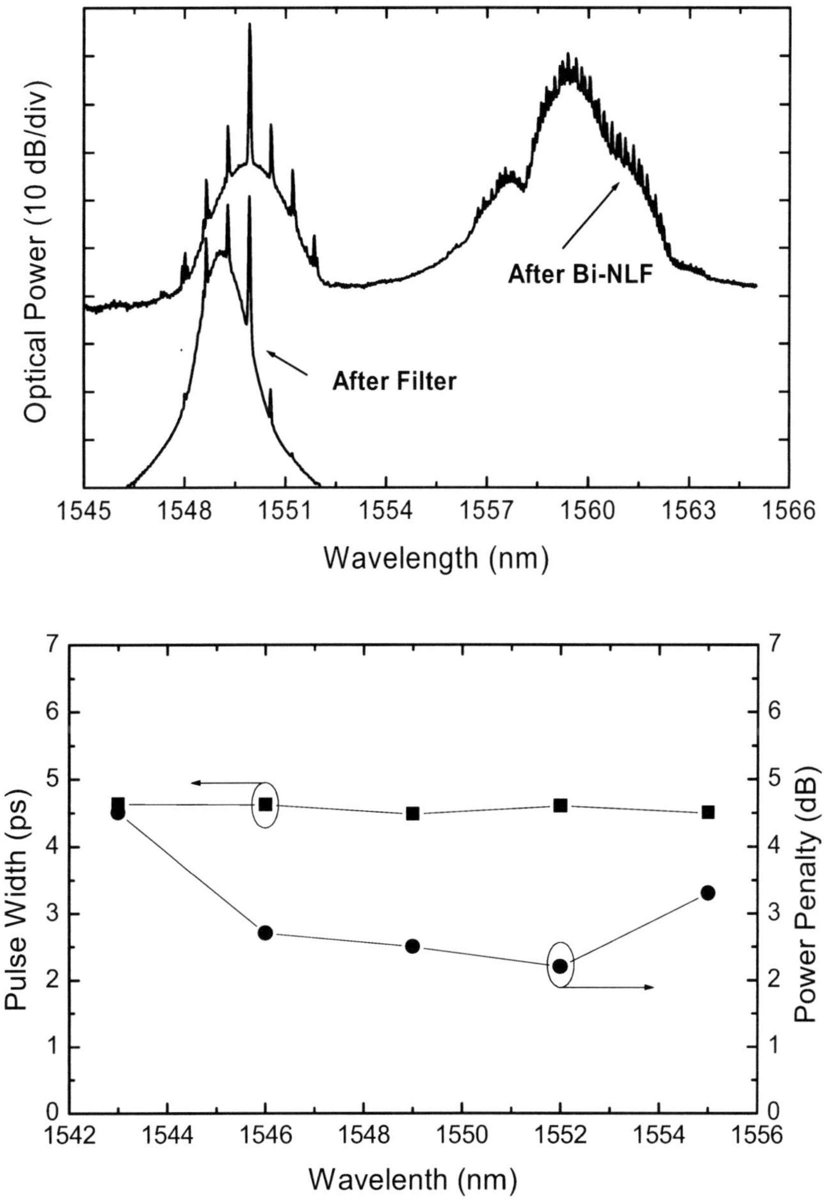

(a)

(b)

Figure 48. (a) Measured optical spectra after the 1-m Bi-NLF. (b) Optical pulse width and power penalty (at BER=10^{-9}) measured as a function of probe wavelength. (after Ref. [118])

5.11. Wavelength Converter and Demultiplexer based on XPM and Subsequent Spectral Filtering

The use of short length of Bi-NLF for the implementation of nonlinear signal processing devices based on XPM and subsequent narrow band offset spectral filtering has also been demonstrated [118]. Figure 47 shows an experimental schematic for wavelength conversion and demultiplexing based on the XPM and subsequent offset filtering principle in short lengths of Bi-NLF.

The 80-Gbit/s input OTDM data stream and the probe beam were combined by a 50:50 coupler and launched into a 1-m-long Bi-NLF. The wavelength-converted signal was obtained through a 1-nm; double-cavity bandpass filter was employed with a 1-nm offset relative to the probe wavelength at the output of the Bi-NLF. For the 80-to-10 Gbit/s demultiplexing, wavelength blue shift of the data pulses induced by XPM from the control pulse trailing edge in a 2-m long Bi-NLF was employed [77]. A 0.6-nm double-cavity bandpass filter was located at the output of the Bi-NLF with a 1.5-nm offset relative to the 80-Gbit/s converted signal wavelength for obtaining the wavelength-blueshifted components only.

The measured output optical spectra are shown in Figure 48(a). The CW probe beam is readily converted into an 80-Gbit/s signal due to XPM induced by the input signal. An error-free wavelength tuning range of ~10 nm was obtained as shown in Figure 48(b), which was limited by both lower XPM efficiency at shorter wavelengths induced by non-negligible walk-off effect and inter-channel interference noise caused by the broadened temporal width of the converted pulses.

CONCLUSION

The state-of-the-art nonlinear optical fiber technologies for the implementation of all-optical nonlinear signal processing devices have been reviewed. In particular, the focus has been given to Bismuth oxide based nonlinear optical fiber and photonic crystal fiber technologies. A huge drawback of optical fiber based nonlinear signal processing devices relative to the semiconductor based ones has always been the stability and compact issues due to its long fiber length. However, the use of short length of ultra-highly nonlinear fiber could readily sort out the problems. It is noticeable that the whole device length in the case of Bi-NLF based devices is 1-m or less, which is almost comparable to that of the devices based on fiber pigtailed semiconductors. This represents a considerable advance in the field of nonlinear fiber optics. The works described above represent a small sample of the body of the works associated with high nonlinearity optical fiber based optical signal processing. There are numerous other works, which are not described in this article.

However, further study is still required to investigate issues such as coupling efficiency, polarization sensitivity, loss, and optimization of dispersion characteristics associated with the use of such devices in real telecommunication or photonic systems. Ultra-high nonlinearity fiber technologies offer significant advantages in terms of reduced device length, reduced power requirements for a variety of nonlinear optical devices. Ultimately, ultra-high nonlinearity fiber technologies should prove to be a powerful way to realize a wide range of practical nonlinear optical devices for fiber-optic communication systems.

REFERENCES

[1] B.-E. Olsson, P. Ohlen, L. Rau, and D. J. Blumenthal, "A simple and robust 40-Gb/s wavelength converter using fiber cross-phase modulation and optical filtering," *IEEE Photon. Technol. Lett.*, vol.12, no.7, pp. 846-848, 2000.

[2] J. Yu, X. Zheng, C. Peucheret, A. T. Clausen, H. N. Poulsen, and P. Jeppesen, "40-Gb/s all-optical wavelength conversion based on a nonlinear optical loop mirror," *J. Lightwave Technol.*, vol.18, no.7, pp.1001-1006, 2000.

[3] D'Ottavi, F. Girardin, L. Graziani, F. Martelli, P. Spano, A. Mecozzi, S. Scotti, R. Dall'Ara, J. Eckner, and G. Guekos, "Four-wave mixing in semiconductor optical amplifiers: a practical tool for wavelength conversion," *IEEE J. Select. Topics Quantum Electron.*, vol.3, no.2, pp.522-528, 1997.

[4] T. Yamamoto, E. Yoshida, and M. Nakazawa, "Ultrafast nonlinear optical loop mirror for demultiplexing 640Gbit/s TDM signals," *Electron Lett.*, vol.34, no.10, pp.1013-1014, 1998.

[5] I. D. Phillips, A. D. Ellis, H. J. Thiele, R. J. Manning, and A. E. Kelly, " 40Gb/s all-optical data regeneration and demultiplexing with long pattern lengths using a semiconductor nonlinear interferometer," *Electron Lett.*, vol.34, no.24, pp.2340-2342, 1998.

[6] Y. Khrushchev, I. D. Phillips, A. D. Ellis, R. J. Manning, D. Nesset, D. G. Moodie, R. V. Penty, and I. H. White, "OTDM applications of dispersion imbalanced fiber loop mirror," *Electron Lett.*, vol.35, no.14, pp.1183-1185, 1999.

[7] J. H. Lee, P. C. Teh, P. Petropoulos, M. Ibsen and D. J. Richardson, "Reduction of interchannel interference noise in a two-channel, grating

based OCDMA system using a nonlinear optical loop mirror," *IEEE Photon. Technol. Lett*, vol.13, no.5, pp.529-531, 2001.

[8] Y. Emori, and S. Namiki, "100nm bandwidth flat gain Raman amplifiers pumped and gain-equalized by 12-wavelength-channel WDM high power laser diodes," in *Proc. Optical Fiber Communications Conference (OFC'99)*, March 1999, Postdeadline paper PD19.

[9] P. C. Reeves-Hall, D. A. Chestnut, C. J. S. De Matos, and J. R. Taylor, "Dial wavelength pumped L- and U-band Raman amplifier," *Electron. Lett.*, vol.37, no.14, pp.883-884, 2001.

[10] D. M. Patrick, and R. J. Manning, "20Gbit/s all-optical clock recovery using semiconductor nonlinearity," *Electron. Lett.*, vol.30, no.2, pp.151-152, 1994.

[11] P. V. Mamyshev, "All-optical data regeneration based on self-phase modulation," in *Proc. European Conference on Optical communication (ECOC'98)*, pp.475-476, 1998.

[12] Schubert, J. Berger, S. Diez, H. J. Ehrke, R. Ludwig, U. Feiste, C. Schmidt, H. G. Weber, G. Totchiyski, S. Randel, and K. Petermann, " Comparison of interferometric all-optical switches for demultiplexing applications in high-speed OTDM systems," *J. Lightwave Technol.,* vol.20, pp.618-624, 2002.

[13] T. Miyazaki, F. Kubota, "All-optical reshaping in a 160-Gb/s OTDM receiver using an XAM gate followed by an SPM discriminator," *IEEE Photon. Technol. Lett.*, vol.16, pp.1909-1911, 2004.

[14] N. G. R. Broderick, T. M. Monro, P. J. Bennett, and D. J. Richardson, "Nonlinearity in holey optical fibers: measurement and future opportunities", *Opt. Lett.*, vol.24, no.20, pp.1395-1397, 1999.

[15] N. Sugimoto, T. Nagashima, T. Hasegawa, S. Ohara, K. Taira, and K. Kikuchi, "Bismuth-based optical fiber with nonlinear coefficient of 1360 $W^{-1}km^{-1}$," in *Proc. Optical Fiber Communications Conference (OFC'2004),* Anaheim USA, March 2004, PDP26 (Postdeadline paper).

[16] M. Jinno, T. Sakamoto, J.-I. Kani; S. Aisawa, K. Oda, M. Fukui, H. Ono, M. Yamada, and K. Oguchi, "1580-nm band, equally spaced 8X10 Gb/s WDM channel transmission over 360 km (3X120 km) of dispersion-shifted fiber avoiding FWM impairment," *IEEE Photon. Technol. Lett.*, vol.10, no.3, pp.454-456, 1998.

[17] P. M. Kjeldsen, M. Obro, J. S. Madsen, and S. K. Nielsen, "SRS induced depletion of 1540nm signal copropagating with 1630nm OTDR pulses," *Electron. Lett.*, vol.32, no.20, pp.1914-1916, 1996.

[18] Y. R. Shen, *The Principles of Nonlinear Optics* (John Wiley & Sons, 1984).

[19] R. W. Boyd, *Nonlinear Optics* (Academic Press, Inc., 1992).

[20] G. P. Agrawal, *Nonlinear Fiber Optics*, (Academic Press, Inc., Third Edition, 2001).

[21] N. J. Doran, and D. Wood, "Soliton processing element for all-optical switching and logic," *J. Opt. Soc. Am. B*, vol.4, no.11, 1843, 1987.

[22] P. Petropoulos, T. M. Monro, W. Belardi, K. Furusawa, J. H. Lee, and D. J. Richardson, "A 2R-regenerative all-optical switch based on a highly nonlinear fiber", *Opt. Lett.*, vol. 26, no.16, pp.1233-1235, 2001.

[23] M. Asobe, T. Kanamori, and K. Kubodera, "Ultrafast all-optical switching using highly nonlinear Chalcogenide glass fiber," *IEEE Photon. Technol. Lett.*, vol.4, no.4 , pp.362-365, 1992

[24] K. Inoue, and H. Toba, "Wavelength conversion experiment using fiber four-wave mixing," *IEEE Photon. Technol. Lett.*, vol.4, no.1 , pp.69-72, 1992.

[25] A. T. Clausen, L. Oxenlowe, C. Peucheret, H. N. Poulsen, P. Jeppesen, S. N. Knudsen, and L. Gruner-Nielsen, "10-GHz return-to-zero pulse source tunable in wavelength with a single- or multiwavelength output based on four-wave mixing in a newly developed highly nonlinear fiber", *IEEE Photon. Technol. Lett.*, vol.13, no.1 , pp.70-72, 2001.

[26] S. Watanabe, T. Naito, and T. Chikama, "Compensation of chromatic dispersion in a single-mode fiber by optical phase conjugation", *IEEE Photon. Technol. Lett.*, vol.5, no.1 , pp.92-94, 1993.

[27] S. T. Davey, D. L. Williams, B. J. Ainslie, W. J. M. Rothwell, and B. Wakefield, "Optical gain spectrum of GeO_2-SiO_2 Raman fiber amplifiers," *IEE Proc.-J*, vol136, no.6, pp.301-306, 1989.

[28] M. N. Islam, "Raman amplifiers for telecommunications," *IEEE J. Sel. Quantum Electron.*, vol.8, no.3, pp.548-559, 2002.

[29] R. G. Smith, "Optical power handling capacity of low loss optical fibers as determined by stimulated Raman and Brillouin scattering", *Appl. Optics*, vol.11, no.11, pp.2489-2494, 1972.

[30] V. E. Perlin, and H. G. Winful, "Distributed feedback fiber Raman laser," *IEEE J. Quantum Electron.*, vol.37, no.1, pp.38-47, 2001.

[31] R. M. Shelby, M. D. Levenson, and P. W. Bayer, "Resolved forward Brillouin scattering in optical fibers", *Phys. Rev. Lett.*, vol.54, no.9, pp.939-942, 1985.

[32] L. Tang, "Saturation and spectral characteristics of the Stokes emission in the stimulated Brillouin process", *J. Appl. Phys.*, vol.37, no.8, pp.2945-2955, 1966.

[33] Y. Tanaka, and K. Hotate, "Analysis of fiber Brillouin ring laser composed of single-polarization single-mode fiber," *J. Lightwave Technol.* Vol.l5, no.5, pp.838-844, 1997.

[34] T. Okuno, M. Tanaka, M. Hirano, T. Kato, M. Shigematsu, and M. Onishi, "Highly nonlinear and perfectly dispersion-flattened fibers for quasi-tunable wavelength conversion," in *Proc. European Conference on Optical Communication (ECOC'2003)*, Rimini Italy, September 1998, vol.3, 614-615.

[35] J. H. Lee, W. Belardi, K. Furusawa, P. Petropoulos, Z. Yusoff, T. M. Monro, and D. J. Richardson, "Four-wave mixing based, 10 Gbit/s tuneable wavelength conversion using a holey fiber with a high SBS threshold," *IEEE Photon. Technol. Lett.*, vol.15, pp.440-442, 2003.

[36] T. M. Monro, K. M. Kiang, J. H. Lee, K. Frampton, Z. Yusoff, R. Moore, J. Tucknott, D. W. Hewak, H. N. Rutt, and D. J. Richardson, "High nonlinearity extruded single-mode holey optical fibers," in *Proc. Optical Fiber Communications Conference (OFC'2002),* Anaheim USA, March 2002, FA1 (Postdeadline paper).

[37] H. Ebendorff-Heidepriem, P. Petropoulos, S. Asimakis, V. Finazzi, R. Moore, K. Frampton, F. Koizumi, D. Richardson, and T. Monro, "Bismuth glass holey fibers with high nonlinearity," *Opt. Exp.*, vol. 12, no. 21, pp. 5082-5087, 2004.

[38] K. K. Y. Wong, M. E. Marhic, and L. G. Kazovsky,, "Phase-conjugate pump dithering for high-quality idler generation in a fiber optical parametric amplifier," *IEEE Photon. Technol. Lett.,* vol.15, pp.33-35, 2003.

[39] J. H. Lee, T. Tanemura, T. Nagashima, T. Hasegawa, S. Ohara, N. Sugimoto, and K. Kikuchi, "Experimental Comparison of Kerr Nonlinearity Figure-of-Merit Including Stimulated Brillouin Scattering Threshold for State-of-the-Art Nonlinear Optical Fibers," *Opt. Lett.,* vol.30, no.13, pp.1698-1700, July 2005.

[40] J. H. Lee, T. Nagashima, T. Hasegawa, S. Ohara, N. Sugimoto, and K. Kikuchi, "Bismuth oxide-based nonlinear fiber with a high SBS threshold and its application to four-wave mixing wavelength conversion of 40-Gbit/s NRZ signal using pure continuous-wave pump," *J. Lightwave Technol.*, vol.24, no.1, pp.22-28, January 2006.

[41] A. Boskovic, S. V. Chernikov, J. R. Taylor, L. Gruner-Nielsen, and O. A. Levring, "Direct continuous measurement of n_2 in various types of telecommunication fiber at 1.55μm," *Opt. Lett.*, vol.21, pp.1966-1968, 1996.

[42] T. Tanemura, Y. Takushima, and K. Kikuchi, "Narrowband optical filter, with a variable transmission spectrum, using stimulated Brillouin scattering in optical fiber," *Opt. Lett.*, vol.27, pp.1552-1554, 2002.

[43] A. Yeniay, J. -M. Delavaux, and J. Toulouse, "Spontaneous and stimulated Brillouin scattering gain spectra in optical fibers, *J. Lightwave Technol.*, vol.20, pp.1425-1432, 2002.

[44] P. Russell, "Photonic crystal fibers," Science, vol.29, pp.358-362, 2003.

[45] T. A. Birks, J. C. Knigth, and P. St. J. Russell, "Endlessly single-mode photonic crystal fiber," *Opt. Lett.*, vol.22, no.13, pp.96-963, 1997.

[46] T. M. Monro, D.J. Richardson, N.G.R. Broderick, and P.J. Bennett, "Holey optical fibers: an efficient modal model," *J. Lightwave Technol.*, vol.17, no.6, pp.1093-1102, 1999.

[47] G. P. Agrawal, *Fiber optic communication systems* (Wiley Interscience, Second Edition 1997).

[48] J. E. Sharping, M. Fiorentino, P. Kumar, and R. S. Windeler, "All-optical switching in microstructured fiber," *IEEE Photon. Technol. Lett.*, vol.14, no.1, pp.77-79, 2002.

[49] T. M. Monro, Y. D. West, D. W. Hewak, N. G. R. Broderick, and D. J. Richardson, "Chalcogenide holey fibers," *Electron. Lett.,* vol.36, no.24, pp.1998-2000, 2000.

[50] P. Petropoulos, T. M. Monro, H. Ebendorff-Heidepriem, K. Frampton, R. C. Moore, H. N. Rutt, D. J. Richardson, "Soliton-self-frequency-shift effects and pulse compression in an anomalously dispersive high nonlinearity lead silicate holey fiber," *Proc. Optical Fiber Communications Conference (OFC'2003)*, Atlanta USA, Postdeadline paper PD3, March 2003.

[51] J. C. Knight et al., "Anomalous dispersion in photonic crystal fiber," *IEEE Photon Technol. Lett.*, vol.12, no.7, pp.807-809, 2000.

[52] W. J. Wadsworth, J. C. Knight, A. Ortigosa-Blanch, J. Arriaga, E. Silvestre, and P. St. J. Russell, "Soliton effects in photonic crystal fibers at 850 nm," *Electron. Lett.*, vol.36, no.1, pp.53-55, 2000.

[53] J. H. V. Price, K. Furusawa, T. M. Monro, L. Lefort, and D. J. Richardson, "A tuneable femtosecond pulse source operating in the range 1.06-1.33 microns based on an Yb doped holey fiber amplifier", in *Proc. Conference*

on Lasers and Electro-Optics (CLEO'2001), Postdeadline CPD1-1, May 2001.

[54] T. A. Birks, D. Mogilevtsev, J. C. Knight, and P. St. J. Russell, "Dispersion compensation using single-material fibers," *IEEE Photon. Technol. Lett*, vol.11, no.6, pp.674-677, 1999.

[55] A. Ferrando, E. Silvestre, J. J. Miret, J. A. Monsoriu, M. V. Andres, and P. St. J. Russell, "Designing a photonic crystal fiber with flattened chromatic dispersion*,*" *Electron. Lett.*, vol.35, no.4, pp.325-327, 1999.

[56] K. Uesaka, K. K.-Y. Wong, M. E. Marhic, and L. G. Kazovsky, "Wavelength exchange in a highly nonlinear dispersion-shifted fiber: Theory and experiments" *IEEE J. Sel. Quantum Electron.*, vol.8, pp.560-568, 2002.

[57] J. H. Lee, P. C. Teh, Z. Yusoff, M. Ibsen, W. Belardi, T. M. Monro, and D. J. Richardson, "A holey fiber based nonlinear thresholding device for optical CDMA receiver performance enhancement," *IEEE Photon. Technol. Lett.*, vol.14, no.6, pp.876-878, June 2002.

[58] S. H. Chang, S. K. Kim, M.-J. Chu, and J.-H. Lee, "Limitations in fiber Raman amplifiers imposed by Rayleigh scattering of signals," *Electron. Lett.*, vol.38, no.16, pp.865-867, 2002.

[59] Z. Yusoff, J.H. Lee, W. Belardi, T.M. Monro, P.C. Teh, and D.J. Richardson, "Raman effects in a highly nonlinear holey fiber: amplification and modulation," *Opt. Lett.,* vol.27, no.6, pp.424-426, March 2002.

[60] G. Burdgel, S. Alam, A. Grudinin, I. Khrushchev , M. Durkin, M. Ibsen, and I. White, "Ultrafast Intensity Modulation via Raman Gain for All-Optical In-Fiber Processing," *Opt.. Lett.*, vol.23, no.8, pp.606-608, 1998.

[61] J. H. Lee, Z. Yusoff, W. Belardi, M. Ibsen, T. M. Monro, and D. J. Richardson, "A tuneable WDM wavelength converter based on cross phase modulation effects in holey fiber," *IEEE Photon. Technol. Lett.,* vol.15, March 2003.

[62] O. Aso, S. Arai, T. Yagi, M. Tadakuma, Y. Suzuki, and S. Namiki, "Broadband four-wave mixing generation in short optical fibre," *Electron. Lett.*, v.36, p.709-711, 2000.

[63] J. E. Sharping, M. Fiorentino, A. Coker, P. Kumar, and R. S. Windeler, "Four-wave mixing in microstructure fiber," *Opt. Lett.*, vol.26, pp.1048-1050, 2001.

[64] J. Hansryd, F. Dross, M. Westlund, P. A. Antrekson, and S. N. Knudsen, "Increase of the SBS threshold in a short highly nonlinear fiber by

applying a temperature distribution," *J. Lightwave Technol.,* vol.19, pp.1691-1697, 2001.

[65] J. H. Lee, Z. Yusoff, W. Belardi, M. Ibsen, T. M. Monro, B. Thomsen, and D. J. Richardson, "Investigation of Brillouin effects in small-core holey optical fiber: lasing and scattering," *Opt. Lett.*, vol.27, pp.927-929, 2002.

[66] L. K. Oxenløwe, A. I. Siahlo, K. S. Berg, A. T. Clausen, B. M. Sørensen, K. Yvind, P. Jeppesen, K. P. Hansen, K. Hoppe, and J. Hanberg, "A novel 160 Gb/s receiver configuration including a glass crystal pulsed laser, photonic crystal fibre and a simple dynamic clock recovery scheme," *Proc. 29th European Conference on Optical Communication (ECOC'2003),* Rimini Italy, Th2.5.3, September 2001.

[67] T. Nagashima. T. Hasegawa, H. Ohara, and N. Sugimoto, "Dispersion Shifted Bi_2O_3-based Photonic Crystal Fiber," in *Proc. European Conference on Optical Communication (ECOC'2006),* Cannes France, paper We.1.3.2, 2006.

[68] X. Feng, T. Monro, P. Petropoulos, V. Finazzi, and D. Hewak, "Solid microstructured optical fiber," *Opt. Exp.*, vol. 11, no.18, pp.2225-2230, 2003.

[69] V.V. Ravi Kumar, A. George, W. Reeves, J. Knight, P. Russell, F. Omenetto, and A. Taylor, "Extruded soft glass photonic crystal fiber for ultrabroad supercontinuum generation," *Opt. Exp.*, vol. 10, no.25, pp.1520-1525, 2005.

[70] S. Ohara, N. Sugimoto, K. Ochiai, H. Hayashi, Y. Furusawa, T. Hirose, T. Nagashima, and M. Reyes, "Ultra-wideband amplifiers based on Bi_2O_3-EDFAs," *Optical Fiber Technol.*, vol.10, pp.283-295, 2004.

[71] K. Kikuchi and K. Taira, "Highly nonlinear bismuth oxide-based glass fibres for all-optical signal processing," *Electron. Lett.,* vol. 38, no. 4, pp.166 – 167, 2002.

[72] N. Nakazawa, H. Kubota, and K. Tamura, "Random evolution and coherence degradation of a high-order optical soliton train in the presence of noise," *Opt. Lett.*, vol.24, pp.318-320, 1999.

[73] Cantini, K. Z. Abedin, G. Sacchi, F. D. Pasquale, and F. Kubota, "Measurement of Raman gain coefficient in Bismuth-based single mode optical fibers" in *Proc. 10th Optoelectronics and Communication Conference (OECC'2005)*, Seoul Republic of Korea, paper 6D2-5, July 2005.

[74] K. K. Y. Wong, K. Shimizu, M. E. Marhic, K. Uesaka, G. Kalogerakis, and L. G. Kazovsky, "Continuous-wave fiber optical parametric

wavelength converter with +40-dB conversion efficiency and a 3.8-dB noise figure, *Opt. Lett.,* vol.28, pp.692-694, 2003.

[75] J. H. Lee, T. Nagashima, T. Hasegawa, S. Ohara, N. Sugimoto, and K. Kikuchi, "Four-Wave Mixing Based Wavelength Conversion of 40-Gbit/s Non-Return-to-Zero Signal Using 40-cm Bismuth Nonlinear Optical Fiber," *IEEE Photon. Technol. Lett,* vol.17, no.7, pp.1474-1476, July 2005.

[76] J. H. Lee, T. Tanemura, T. Nagashima, T. Hasegawa, S. Ohara, N. Sugimoto, and K. Kikuchi, "Use of 1-m Bi_2O_3 Nonlinear Fiber for 160-Gbit/s Optical-Time Division Demultiplexing Based on Polarization Rotation and Wavelength Shift Induced by Cross-Phase Modulation," *Optics Letters*, vol.30, no.11, pp.1267-1269, June 2005.

[77] J. Li, B. E. Olsson, M. Karlsson, and P. A. Andrekson, "OTDM demultiplexer based on XPM-induced wavelength shifting in highly nonlinear fiber," *IEEE Photon. Technol. Lett,* vol.15, no.12, pp.1770-1772, 2003.

[78] L. K. Oxenløwe, L. J. Christiansen, D. Larsson, K. Yvind, A. T. Clausen, J. Seoane, A. I. Siahlo, B. Sørensen, and P. Jeppesen "Pre-scaled clock recovery with compact semiconductor devices for ultra high-speed OTDM systems," *29th European Conference on Optical Communication (ECOC'2003),* Rimini Italy, paper Th4-1, September 2003.

[79] Zibar, J. Mørk, L. K. Oxenløwe, M. Galili, and A. T. Clausen, "Timing jitter analysis for clock recovery circuits based on an optoelectronic phase-locked loop (OPLL)," *Proc. Conference on Laser and Electro-Optics (CLEO'2005),* Baltimore USA, paper CMZ4, June 2005.

[80] J. H. Lee, S. Ohara, T. Nagashima, T. Hasegawa, N. Sugimoto, K. Igarashi, K. Katoh, and K. Kikuchi, "Clock Recovery and Demultiplexing of High-Speed OTDM Signal through Combined Use of Bismuth Oxide Nonlinear Fiber and Erbium-Doped Bismuth Oxide Fiber," *IEEE Photon. Technol. Lett.*, vol.17, no.12, pp.2658-2660, December 2005.

[81] J. H. Lee, T. Nagashima, T. Hasegawa, S. Ohara, N. Sugimoto, and K. Kikuchi, "Wavelength Conversion of 160 Gbit/s OTDM Signal Using 1-m-long Bismuth Oxide-based Highly-Nonlinear Fiber in the Kerr Shutter Configuration," *Electron. Lett.,* vol.41, no.16, pp.918-919, August 2005.

[82] J. H. Lee, T. Nagashima, T. Hasegawa, S. Ohara, N. Sugimoto, and K. Kikuchi, "All-fiber based 80-Gbit/s wavelength converter using 1-m long Bismuth nonlinear optical fiber with a nonlinearity γ of 1100 $W^{-1}km^{-1}$," *Opt. Exp.,* vol.13, no.8, pp.3144-3149, April 2005.

[83] G.-W. Lu, L.-K. Chen, C.-K. Chan, and C. Lin, "All-optical tunable wavelength conversion based on cross-polarisation modulation in nonlinear photonic crystal fibre," *Electron. Lett.* vol.41, pp.55-56, 2005.

[84] T. Morioka, H. Takara, K. Mori, and M. Saruwatari, "Ultrafast reflective optical Kerr demultiplexer using polarization rotation mirror," *Electron. Lett.*, vol.28, pp.521-522, 1992.

[85] C. Schubert, C. Schmidt, S. Ferber, R. Ludwig, and H. G. Weber, "Error-free all-optical add-drop multiplexing at 160 Gbit/s," *Electron. Lett.*, vol.39, pp.1074-1076, 2003.

[86] H.-F. Chou, J. E. Bowers, and D. J. Blumenthal, "Compact 160-Gb/s add-drop multiplexer with a 40-Gb/s based rate using electroabsorption modulators," *IEEE Photon. Technol. Lett.*, vol.16, pp.1564-1565, 2004.

[87] J. Suzuki, K. Taira, Y. Ozeki, T. Tanemura, and K. Kikuchi, "All-optical time-division add-drop multiplexer using optical fibre Kerr shutter," *Electron. Lett.*, vol.40, pp.445-446, 2004.

[88] J. H. Lee, T. Tanemura, Y. Takushima and K. Kikuchi, "All-optical 80 Gbit/s add-drop multiplexer using fiber-based nonlinear optical loop mirror," *IEEE Photon. Technol. Lett.* vol.17, pp.840-842, 2005.

[89] L. Rau, S. Rangarajan, W. Wang, and D. J. Blumenthal, "All-optical add-drop of an OTDM channel using an ultra fast fiber based wavelength converter," in *Proc. Optical Fiber Communications Conference (OFC'2002),* Anaheim USA, paper WM1, march 2002.

[90] J. H. Lee, T. Nagashima, T. Hasegawa, S. Ohara, N. Sugimoto, and K. Kikuchi, "All fiber-based 160-Gbit/s add/drop multiplexer incorporating a 1-m-long Bismuth Oxide-based ultra-high nonlinearity fiber," *Opt. Exp.,* vol.13, no.18, pp.6864-6869, September 2005.

[91] D. M. Patrick and A. D. Ellis, "Demultiplexing using cross phase modulation-induced spectral shifts and Kerr polarisation rotation in optical fibre," *Electron. Lett.*, 29, 227- 29 (1993).

[92] B. S. Robinson, S. A. Hamilton, and S. J. Savage, "40 Gbit/s all-optical XOR using a fiber-baser folded ultrafast nonlinear interferometer," in *Proc. Optical Fiber Communications Conference (OFC'2002)*, Anaheim USA, paper ThY2, 2002.

[93] T. A. Ibrahim, R. Grover, L.-C. Kuo, L. C. Calhoun, and P. T. Ho, "All-optical AND/NAND logic gates using semiconductor microresonators," *IEEE photon. Technol. Lett.*, vol.15, pp.1422-1424, 2003.

[94] C. Yu, L. Christen, T. Luo, Y. Wang, Z. Pan, L. Yan, and A. E. Willner, "All-optical XOR gate using polarization rotation in single highly

nonlinear fiber," *IEEE Photon. Technol. Lett.*, vol.17, pp.1232-1234, 2005.

[95] J. H. Lee, T. Nagashima, T. Hasegawa, S. Ohara, N. Sugimoto, and K. Kikuchi, "40-Gbit/s XOR and AND Gates Using Polarisation Switching within 1-m-long Bismuth Oxide-based Nonlinear Fibre," *Electron. Lett.,* vol. 41, no.19, pp.1074-1075, *September* 2005.

[96] A. Chowdhury, G. Raybon, R.-J. Essiambre, J. H. Sinsky, A. Adamiecki, J. Leuthold, C. R. Doerr, and S. Chandrasekhar, "Compensation of intrachannel nonlinearities in 40-Gb/s pseudolinear systems using optical-phase conjugation," *J. Lightwave Technol.*, vol.23, no.1, pp.172-177, 2005.

[97] I. Mecozzi, G. Contestable, F. Martelli, L. Graziani, A. D'Ottavi, P. Spano, R. Dall'Ara, J. Eckner, F. Girardin, and G. Guekos, "Optical spectral inversion without frequency shift by four-wave mixing using two pumps with orthogonal polarization," *IEEE Photon. Technol. Lett,* vol.10, no.3, pp.355-357, 1998.

[98] S. Watanabe, T. Naito, and T. Chikama, "Compensation of chromatic dispersion in a single-mode fiber by optical phase conjugation," *IEEE Photon. Technol. Lett,* vol.5, no.1, pp.92-95, 1993.

[99] C. Peucheret, B. Zsigri, P. A. Andersen, K. S. Berg, A. Tersigni, P. Jeppesen, K. P. Hansen, and M. D. Nielsen, "40 Gbit/s transmission over photonic crystal fibre using mid-span spectral inversion in highly nonlinear photonic crystal fibre," *Electron. Lett.,* vol.39, no.12, pp. 919-921, 2003.

[100] M. N. Chou, I. Brener, M. M. Fejer, E. E. Chaban, and S. B. Christman, "1.5μm band wavelength conversion based on cascaded second-order nonlinearity in LiNbO3 waveguides,". *IEEE Photon. Technol. Lett,* vol.11, no.6, pp.653-655, 1999.

[101] S. L. Jansen, D. van den Borne, C. Climent Monsalve, S. Spalter, P. M. Krummrich, G. D. Khoe, and H. de Waardt, "Reduction of Gordon-Mollenauer phase noise by midlink spectral inversion," *IEEE Photon. Technol. Lett.,* vol.17, no.4, pp.923-925, 2005.

[102] M. Tani, and S. Yamashita, "Dispersion compensation with SBS-suppressed fibre phase conjugator using synchronized phase modulation," *Electron. Lett.,* vol.39, no.19, pp. 1375-1377, 2003.

[103] J. H. Lee, C. H. Kim, Y.-G. Han, S. B. Lee, T. Nagashima, T. Hasegawa, S. Ohara, N. Sugimoto, and K. Kikuchi, "Bismuth Nonlinear Fibre Based Optical Phase Conjugator without SBS-Induced Efficiency Limitation and

Its Application to Dispersion Compensation in a Transmission Link," *Electron. Lett.*, vol.42, no.5, pp.298-299, March 2006.

[104] I. Bornholdt, J. Slovak, and B. Sartorius, "Semiconductor-based all-optical 3R regenerator demonstrated at 40 Gbit/s," *Electron. Lett.*, vol. 40, no.3, pp.192-194, 2004.

[105] T. Otani, T. Miyazaki, and S. Yamamoto, "40-Gb/s optical 3R regenerator using electroabsorption modulators for optical networks," *J. Lightwave Technol.*, vol.20, no.2, pp.195-200, 2002.

[106] J. Suzuki, T. Tanemura, K. Taira, Y. Ozeki, and K. Kikuchi, "All-optical regenerator using wavelength shift induced by cross-phase modulation in highly nonlinear dispersion-shifted fiber," *IEEE Photon. Technol. Lett*, vol.17, no.2, pp.423-425, 2005.

[107] S.-H. Her, G. Raybon, and C. Headley, "Optimization of pulse regeneration at 40 Gb/s based on spectral filtering of self-phase modulation in fiber" *IEEE Photon. Technol. Lett.*, vol.16, no.1 pp.200-202, 2004.

[108] Raybon, Y. Su, J. Leuthold, R.-J. Essiambre, T. Her, C. Joergensen, P. Stenvurzel, K. Dreyer, and K. Feder, "40 Gbit/s pseudo-linear transmission over one million kilometres," in *Proc. Optical Fiber Communication Conf. (OFC2002),* Anaheim CA, Mar. 17-22, 2002, FD10.

[109] X. Liu, C. Xu, and W. H. Knox, "Characteristics of all-optical 2R regeneration based on self-phase modulation in highly-nonlinear fibers," in *Proc. Conference on Lasers and Electro-Optics (CLEO'2002),* Long Beach USA, CThAA5, 2002,.

[110] L. B. Fu, M. Rochette, V. G. Ta'eed, D. J. Moss, and B. J. Eggleton, "Investigation of self-phase modulation based optical regeneration in single mode As_2Se_3 chalcogenide glass fiber," *Opt. Exp.,* vol.13, no.9, pp.7637-7644, 2005.

[111] J. H. Lee, T. Nagashima, T. Hasegawa, S. Ohara, N. Sugimoto, Y.-G. Han, S. B. Lee, and K. Kikuchi, "Output Performance Investigation of Self-Phase Modulation Based 2R Regenerator Using Bismuth Oxide Nonlinear Fiber," *IEEE Photon. Technol. Lett.*, vol.18, no.12, pp.1296-1298, June 2006.

[112] N. Bergano, F. W. Kerfoot, and C. R., Richardson, "Margin measurements in optical amplifier systems," *IEEE Photon. Technol. Lett.,* vol.5, no.3, pp.304-306, 1993.

[113] J. Yu, X. Zheng, C. Peucheret, A. T. Clausen, H. N. Poulsen, and P. Jeppensen, "All-optical wavelength conversion of short pulses and NRZ

signals based on a nonlinear optical loop mirror" *J. Lightwave Technol.*, vol.18, pp.1007-1017, 2000.

[114] T. Sakamoto and K. Kikuchi, "Nonlinear optical loop mirror with an optical bias controller for achieving full-swing operation of gate switching," *IEEE Photon. Technol. Lett.,* vol.16, no.2, pp.545-547, 2004.

[115] T. Tanemura, H. C. Lim, K. Kikuchi, "Suppression of idler spectral broadening in highly efficient fiber four-wave mixing by binary-phase-shift-keying modulation of pump wave," *IEEE Photon. Technol. Lett.*, vol.13, no.12, pp.1328-1330, 2001.

[116] J. H. Lee, T. Nagashima, T. Hasegawa, S. Ohara, N. Sugimoto, and K. Kikuchi, "Wideband Tunable Wavelength Conversion of 10-Gbit/s Non-Return-to-Zero Signal Using Cross Phase Modulation-induced Polarization Rotation in 1-m Bismuth Oxide-based Nonlinear Optical Fiber," *IEEE Photon. Technol. Lett.*, vol.18, no.1, pp.298-300, January 2006.

[117] L. Möller, S. Yikai, L. Xiang, J. Leuthold, and X. Chongjin, "Ultrahigh-speed optical phase correlated data signals," *IEEE Photon. Technol. Lett.* vol.15, pp.1597-1599, 2003.

[118] J. H. Lee, T. Nagashima, T. Hasegawa, S. Ohara, N. Sugimoto, and K. Kikuchi, "Bismuth Oxide Nonlinear Fibre Based 80-Gbit/s Wavelength Conversion and Demultiplexing Using the Cross Phase Modulation and Filtering Scheme", *Electron. Lett.*, vo.41, no.22, pp.1237-1238, October 2005.

INDEX

A

B

C

D

T

U

V

W